E
O
F
W
Calbourn
Rowlodge
Bowcomb Down
Bowcomb F.
Gallebery Down
Roughborough Down
Lemerston Down
Gatcomb Park
Gatcomb
Gatcomb Down
Chillerton Down
Chillerton
Brixton
Shorwell
North Court
West Court
Yafford
Thorncross
Sutton
Kingston
Billingham Ho.
Dungewood
Lit. Atherfield
Atherfield Gr.
Atherfield F.
Cowleaze Chine
Atherfield Rocks
Atherfield Ledge
Fishing Cove
Whale Chine
Walpen
Walpen Chine
C H A L E
B A Y
Blackgan Chine
Chale
Chale Street
Stroud Green
North Ground
Upper Appleford
Appleford F.
Holbrooks
Niton Down
St. Katherine's Beacon Light Ho.
Chale Down
Rocken End
Buddle
Puckaster Cove
5
10

The Lifeboats of Brighstone Bay

by

CHRISTOPHER J. WILLIS
and
EDWARD H. ROBERTS

ISLE OF WIGHT COUNTY PRESS

THE LIFEBOATS OF BRIGHSTONE BAY

by

Christopher J. Willis
and
Edward H. Roberts

First published by the Association of Old Worcesters 1982
as a contribution to Maritime England Year
under the title 'The Worcester Cadet'
'The Story of a Life-boat'

2nd Edition 1985

ISBN 0 9510379 3 5

PRINTED AND PUBLISHED BY
THE ISLE OF WIGHT COUNTY PRESS
29 HIGH STREET, NEWPORT
ISLE OF WIGHT PO30 1ST

CONTENTS

"The British People are peculiar in many ways, and British seamen and fishermen are the most peculiar of all. They are capable of almost anything, even what seems to be the impossible, if you will let them do it their own way".

JACK SEELY (1st Lord Mottistone)

Model of the life-boat Worcester Cadet *presented to Commander Henderson-Smith by the R.N.L.I. 1880. Currently on display at the St. Catherine's Quay Exhibition at Blackgang Chine*

FOREWORD

by

Captain The Lord Mottistone, C.B.E., Royal Navy

This account of the pulling and sailing life-boats of the south-west coast of the Isle of Wight and their principal rescues makes thrilling reading.

In the latter days of the *Susan Ashley* at Brook, my father, the first Lord Mottistone, was the coxswain and my elder brother Patrick, later the third Lord Mottistone, was one of the oarsmen. As a youth, I had the privilege of being taken on practice service. Even that experience on a calm day was an exciting event, giving one some understanding of the dangers and difficulties of a launch, often at night, in a storm force south-westerly blow.

Christopher Willis and Ed. Roberts are to be congratulated on skilfully drawing together the various accounts of the brave actions of the life-boat crews and the fortitude of their "helpers" during the 77 years existence of the Brighstone Bay life-boats.

PREFACE

This short history tells the story of the Royal National Life-boat stations in Brighstone Bay in the Isle of Wight. The first stations at Brook and Brighstone Grange were opened in 1860 and the last to close was Brook in March 1937.

Perhaps in the telling of these stories *Worcester Cadet* receives undue prominence, this is because she was the starting point of our researches and as former cadets of the famous training ship *H.M.S. Worcester* we can only crave indulgence of our readers.

In our researches we came across many varieties of spelling. Brighstone has from time to time been known as Brixton and as such appears on many maps and in old guide books. It is only in relatively recent years that the Post Office, exasperated with much South London mail finishing up in the Isle of Wight, insisted that the name Brighstone be used. Similarly Brook appears sometimes with an 'e' and sometimes without. In the interests of consistency we have throughout used the modern spelling of both the villages and where we have found variances in the spelling of names of men and ships we have used the most common.

This is the story of an era when power-driven life-boats were unknown. In their daily lives men relied on their self-taught skill in handling boats propelled by sails and oars in all weather by day and by night, winter or summer. They were a remarkable breed of men, tough and determined to save life and who never failed to come forward to man the boats. They were backed up by men and women helpers who, when the boat had been launched, watched and waited to give further help when needed.

As the story unfolds the reader will see that these boats and crews had adventurous lives, they knew triumph, but as the graves in Brook and Brighstone churchyards remind us, they knew tragedy as well. They were however only some of the many R.N.L.I. boats then and now manned and ready to save life at sea. Boats whose service and the brave men that sail in them have been an honour to the R.N.L.I. We hope that this short account will encourage the readers to make the most practical response he or she can by giving support to the R.N.L.I.

DECEMBER 1984 C.J.W., E.H.R.

ACKNOWLEDGMENTS

We would like to express our sincere thanks to all those that have patiently borne our researches in the various newspaper offices, museums and institutions we have visited. In particular we would like to thank the Isle of Wight County Archivist Mr. Clifford Webster, M.A. and his staff at the Record Office in Newport, the Dabell family of Blackgang Chine for all their help and to Lord Mottistone, Mr. R. H. Buckett of Brighstone, Mr. G. E. Cotton of Yarmouth and Mr. Herbert Morris for their reminiscences and records of their and their respective forbears' services in the crew of the boats. We are indebted to Miss D. M. Brooke for details taken from the Brighstone Parish Record, to Mr. Eric Toogood of Totland for access to his records and photographs, to local residents of Brighstone and Brook and to many others. Our thanks are due to Mr. Frank Taylor for reproducing the old photographs, and to Mr. Peter Oliver of Looe for photographing the model for us, to Mr. Simon Nigh and to several others who loaned old photographs. Lastly but by no means least we would wish to thank Mr. N. F. Stripp, the Rescue Records supervisor and the head-quarters staff of the R.N.L.I. at Poole where we have been permitted access to documents, books and other historical information.

Brighstone Bay looking towards the Needles, Atherfield Point centre of the picture
Photograph W. J. Nigh

BRIGHSTONE BAY 1860

"It is a matter of painful notoriety, that 14 lives were lost by shipwreck at the back of the Island, about three weeks ago. It was stated by the officers of coastguard and by others at the Inquest, that if a life-boat had been at hand, the whole of these lives might have been saved".

These opening words from an appeal letter dated 24th December 1859 signed by Edward McAll, the Rural Dean and Rector of Brighstone, and John Pellew Gaze, the Rector of Brook, sets the scene for the beginning of the life-boat service in Brighstone Bay.

The coastline from the Needles to St. Catherine's Point is a dangerous coast particularly in the days of sail. As the map on the inside of the cover shows, there is a continuous line of ledges and rocks extending well off shore. Sailing ship masters were advised to stand well out to sea when approaching this coast because, caught too close in with no room to tack, they could very soon find themselves on a lee shore with little hope of clawing out to sea. Other sailing vessels coming up Channel and making for the Solent were known to have missed the Needles lighthouse and been driven ashore to the eastward. Even in the early days of steam when methods of navigation were not as accurate as they are today, this coastline was to take its toll of wrecks. These were the dangers that the two clergymen were highlighting when they made their call for the setting up of life-boat stations in Brighstone Bay.

Having heard the report of the coastguard men a letter had at once been written to the Royal National Life-boat Institution calling upon them to take immediate steps to rectify the deficiency. The reply was prompt and encouraging, their General Inspector Captain Ward, R.N., would visit the South Wight and, if he so advised, the R.N.L.I. would defray half the expense of placing life-boats at such stations as he might approve, the cost being estimated at about £350 for each boat.

In the event Captain Ward's report concluded that two life-boats were needed between St. Catherine's and the Needles, one to be at Brook and the other at Grange Chine, the latter not being as far to the eastward as he could have wished, but that being the only place from which there was access by a road to the shore.

So the two Rectors launched their appeal to provide the balance of the cost required to be raised locally. The response was so encouraging that six weeks later on 4th February 1860 they were able to write again saying that the money was forthcoming and also to announce another development.

This further development was that the Royal Victoria Yacht Club had, without previous concert, embarked on a similar scheme to provide a life-boat on the South Wight coast. One of their members, Mr. John Turner-Turner, had a letter published in the Times of London calling for £359 to establish a life-boat station at Brook, personally offering £50 to start the fund. This was to be followed by the Commodore of the R.V.Y.C., Mr. G. Holland-Ackers, sending an

The Life-boat building at Grange Chine in use as a cafe about 1935

The remains of t
foundations of th
Life-boat buildin
in May 1983

appeal circular to all the members and the money was forthcoming. Now that sufficient money had been raised locally, the newly formed Isle of Wight Board in connection with R.N.L.I. felt able to release the Parent Board from their obligation to provide half the costs, thus enabling them to carry out work in localities which were less self-supporting.

Congratulations all round were in order and so the life-boat service came to Brook and Brighstone to be joined 30 years later by a third life-boat at Atherfield.

Brighstone sometimes known as Brixton and Brook or Brooke were in 1860 small village communities in the southern part of the Isle of Wight, midway between St. Catherine's Point and the village of Freshwater and about a mile inland from the sea. The population of the two villages was at that time only small but the villagers, mainly those working on the land and fishermen, were to provide the crew for the new life-boats and the nearby farmers to provide the teams of horses, with whole families including women and children acting as "helpers" to launch and recover the life-boat.

The life-boat house at Brook was a stone building standing on high land, given by Charles Seely, at the head of the chine next to the old cottages. Large double doors at one end allowed the boat on its carriage to be pulled forward before commencing its descent to the beach.

The life-boat house at Brighstone Grange was a brick structure standing just behind the cliff edge where Grange Chine meets the beach. The site of the life-boat house and the right of way access over the private road were held by the R.N.L.I. for 99 years from 25th March 1860, the ground rent being ten shillings per annum. Built to house the first life-boat it had to be widened in 1906 to accommodate a larger boat. The cost of widening the building was some £95 and the work was carried out by the local builder Mr. Buckett, a name famous in the history of the Brighstone Grange Station. The cost was defrayed from a legacy left by a Miss Hyde of Oxford and her name was affixed to the gate to mark her generosity. The slipway and the route to the beach were over the years to cause many problems and constant references can be found in the minute books and visiting Inspector's reports to the need to repair and maintain the access, and when the station was finally closed bad access was to be one of the reasons given.

At Atherfield Point the position chosen for housing the life-boat was on the top of the cliff some 75 feet above the beach. A corrugated iron shed was erected on the top, and on the face of the cliff a series of flat skids or sleepers were fastened over which the boat was lowered by ropes and hauled up again by using a powerful winch. This track ran for 240 feet up an incline of 1 in 3. When the boat reached the shore it was manned and then launched into the sea on roller chocks placed under the keel.

When the call came a gun was fired and the crews would hurry to the life-boat stations together with a beach handling party of villagers. In the case of Brook and Brighstone Grange the carters working in the fields would bring down their horses. The boat would be hauled stern first to the beach by a team of up to eight or ten horses, depending on the weather, and the crew would embark. The boat on its carriage was then taken as far as possible into the sea before being turned by the leading horses, so that the bow was pointing directly at the oncoming waves breaking on the shore.

A successful launch depended on three important factors. First, and most important, the coxswain had to choose exactly the right moment to give the order "Launch". The incoming wave had to be big enough to float the boat but not so large as to overwhelm it, also the area of water had to be large enough to enable the boat to go ahead into deep water. Failure to watch these crucial factors could lead to disaster, such as capsizing or throwing out the crew. Launching on a receding wave could have equally disastrous results. The experience and judgement of the coxswain was therefore all important for a successful launch.

Secondly, the helpers often standing in the sea, had to react instantly and together when the coxswain's cry came to launch by hauling on the lines which shot the boat forward off the carriage into the sea using all their weight and strength. For a good launch 60 or more people were needed.

Thirdly, the ability of ten oarsmen to be ready to pull together with their oars the moment the boat was waterborne was crucial. It was also essential that they made their second stroke before the next oncoming wave hit the bow.

Occasionally an outhaul warp from a fixed mooring to seaward was taken on board, the crew hauled on the rope the moment the boat was waterborne and then got their oars out before throwing the warp overboard. Also on occasions, providing the wind was not dead onshore, the sail would be hoisted to assist the launch.

In the case of the Brook boat there was an added complication. The building of the "new" life-boat road meant that the launch was straight at the rocks. It was therefore necessary immediately after launching to turn westwards and pull parallel to the coast for about 200 yards before turning to seaward through a gap in the rocks. It can be imagined how difficult an operation this could be on a dark and stormy night.

The photograph on page 36 of the *Susan Ashley* entering the water shows clearly the efforts of the men and horses and illustrates the drama of a launch.

When it was necessary to communicate between the stations a man would be sent on horseback with a message, but often he would arrive in a state of collapse having galloped through the storm. When this happened the beach party would look to see if he was wearing a leather belt with a brass token or medallion attached to it. This token was authority for the requested boat to be called out. One of these belts can be seen in the St. Catherine's Quay Exhibition at Blackgang.

The first Rescue *seen at Brighstone Grange Life-boat house on 13th August, 1860*

Left to right: Rev. E. McAll – Rector of Brighstone in the top hat, the coxswain Jam Buckett sitting on the shafts of the carriage, Coastguard Officer J. Crew sitting on th shafts to his right, Mr. Cooke, honorary architect, Rev. John Pellew Gaze, Vicar of Broo and honorary secretary of the Isle of Wight Board and George Wyatt, honorary treasure

BRIGHSTONE GRANGE

THE 'RESCUE'

The first life-boat to be placed on station at Brighstone Grange was named *Rescue*. She was to remain in service for six years before being replaced in 1866 by a similar boat of the same name.

On 13th August 1860, the official party having come round from the similar ceremony at Brook, *Rescue* was launched by Lady Heytesbury who dashing a bottle of wine against the bows of the boat announced in a distinct voice "Success to the Royal Victoria Yacht Club life-boat *Rescue*". This was, so the Isle of Wight Observer reported, to be the signal for spontaneous cheering and waving of hankerchiefs, in testimony of the satisfaction with which the establishment of the station was viewed.

The donors, the Royal Victoria Yacht Club, were presented by the R.N.L.I. with a picture of a life-boat going to a wreck, as a small token of their appreciation of the financial assistance rendered by the club.

In the 20 years that these two boats were on station, there were many calls on their service, some false alarms, some occasions when the boat was launched and stood by a ship in trouble without having to render service. The Rescue Board now exhibited in Brighstone Village Reading Room records however six occasions when service was required and a total of 187 lives were saved. The stories of these rescues more than justifies all the time and effort that went into practice drills and launches and the dedicated work of the crews and shore parties.

2nd April 1862 Barque CEDARINE

The barque *Cedarine,* 308 tons, was the last ship to be constructed in the Bermuda Dockyard at Shelley Bay and was built by J. J. Outerbridge. Commanded by Captain T. Dill she was on her maiden voyage from Bermuda to Portsmouth when she ran ashore and was wrecked some 100 yards east of Brighstone Grange in the early morning of 2nd April 1862. She had on board a crew of 13 and was carrying 191 convicts who had accomplished their period of servitude and were returning home. The weather was hazy and the ship was lying 50 yards from the shore when the life-boat found her and was able to take off 134 of those on board, the remainder being rescued by the coastguard service. The convicts who had endured this unexpected end to their sentences were taken by military escort to Parkhurst from where they were released, but not before, as the story relates, they were well tipsy from drinking all the beer in the local inns. The minutes of the Royal Victoria Yacht Club record receiving the thanks of the R.N.L.I. following the first successful rescue carried out by the boat they had donated.

The *Cedarine* was to become a total loss but her cedar planks were salvaged and used to reline the roof of the chancel of the church of St. Peter and St. Paul at Mottistone which was at that time being restored. Many local buildings were no doubt unofficially repaired at the same time.

A picture of the *Cedarine* in full sail painted by the disabled artist Earl Bailly hangs at the art gallery in Shelley Bay, Bermuda and a copy can be seen in Mottistone Church.

19th January 1864 Schooner THETIS

The schooner *Thetis* of London foundered in heavy weather off St. Catherine's Point in the early hours of 19th January 1864 and her crew of four took to their boat. The distress signals sent up before she foundered having been sighted ashore, the *Rescue* was launched and came up with the ship's life-boat and brought the survivors safely ashore. At their meeting in London on 4th February 1864 the R.N.L.I. granted the sum of £4 10s 0d to the crew of the *Rescue* for their work in saving these four lives.

4th May 1872 Brig L'ETOILE

From an account in the Isle of Wight Observer, 18th May 1872.

The *L'Etoile,* a French brig, 285 tons, F. Chattillier, Captain, from Cette, bound to Riga, laden with salt, ran on the rocks a little to the eastward of Sudmore at 9.30 p.m., on the evening of 3rd May 1872. The night was intensely

James Buckett, Brighstone coxswain
Photograph R. H. Buckett

dark, and a fresh gale was blowing from the south-west with rain and a heavy ground swell. The signal light of the vessel in distress was sighted from two points several miles apart by the coastguards, who hastened to the life-boat stations nearest their respective beats – the one at Brook and the other at Brighstone Grange and fired the signal guns, so that both crews were summoned almost simultaneously. From Grange a messenger was dispatched for horses to Thorncross. With great promptitude the proprietor, Mr. E. Mowbray, with his son, two carters (lanterns in hand) and six horses, arrived at the station, where a full muster of crew (ready life-belted) and helpers (33 men) awaited. The horses being attached, the life-boat, on her transporting carriage, immediately left for the beach, and was conveyed along the shore for a distance of a mile and a quarter, to nearly opposite to where the *L'Etoile* lay stranded. The tide being half-ebb, great difficulty was occasioned in launching, owing to the flatteners over which the carriage had to be backed in order to gain a sufficient depth of water, but eventually perseverance and prompt attention to the command of the coxswain proved successful, and in 20 minutes after the launch the life-boat was near enough to enable the ill-fated crew of eight to escape by a rope on the starboard side; and at 2 a.m., they were safely landed on the shore. The life-boat was then replaced on her carriage and drawn along the shore and up the steep ascent to the boathouse at 3 a.m. On examination, there was found to be slight damage to the foremost air compartment and a bow hook and one shaft of the carriage was broken – the former owing to contact with the wreck, and the latter from the man falling whilst holding up the shaft at the time when the horses were attached by haulup rope and drawing the carriage over the flatteners. The French crew were conducted on their way to Yarmouth, along the cliff to Brook life-boat station, where, in the resuscitating room a fire had been lighted, and creature comforts were soon provided. The Brook boat *George and Anne* arrived at the wreck about ten minutes after the crew had been taken off by the *Rescue*.

3rd February 1873 — S.S. WOODHAM

The Norwegian iron screw steamship *Woodham,* Captain Helgeson, 648 tons of Christiania was on passage with a cargo of coal and soda from South Shields to New York when the main shaft of her engine became disabled. She was taken in tow by a passing steamer but in the heavy seas that were running, she broke her tow no less than four times before being finally left to the mercy of the seas and running ashore on Chilton Ledge on the evening of Sunday 3rd February 1873.

Signals of distress were fired and paraffin was lighted leading the spectactors on shore to think that she was on fire. The *Rescue* was launched and made her way out to the wreck to find that in the meantime some of the crew had taken to the ship's boats. Persuaded by the life-boat crew they abandoned their boats, which were promptly smashed in the heavy seas, and in two journeys were brought to shore. The master and the mate, however, refused to leave but afterwards made signals of distress. By this time the *Rescue* was rehoused and her crew on their way home when a third trip had to be made to take them off. When all were rescued the crew of the *Rescue* were, so the minute book recalls, "severely tried and exhausted". For his gallantry on that night James Buckett the coxswain was awarded the R.N.L.I. silver medal.

4th March 1873 S.S. QUAIL

The steamer *Quail* of Cork, on passage from Liverpool to Rotterdam with a general cargo ran ashore on Atherfield Point in dense fog, on the morning of 4th March 1873. When the fog lifted the *Quail* was seen to be ashore and the *Rescue* was launched. In the meantime 14 of the crew had taken to the boats leaving on board, the captain, a stewardess, the Dutch pilot and five of the crew who were duly brought ashore. The ship's boat had by this time made Freshwater Gate and were brought back overland to Atherstone.

9th February 1879 Barque ALPHEUS MARSHALL

It was Sunday evening during divine service when the life-boat gun was fired, and most of the congregation left the church and rushed to the shore. As soon as the horses arrived the coxswain called over the names and gave out the cork jackets. One of the crew failed to answer to his name, and George Downer offered his services and was accepted by Coxswain Buckett. There was a very heavy sea running, but the boat made a successful launch.

No problems were encountered in getting the crew off the *Alpheus Marshall* which had been wrecked on Atherfield Ledge and some hours later the green light was shown, and the shore helpers quickly got ready to help the boat ashore. The helper whose duty it was, if possible, hooked the bent bar attached to the end of a long rope into the hole in the keel of the boat before it touched the beach, so that each wave would help the boat up the beach. He had a rope tied round his waist in case he was knocked down. He failed to catch the boat and the next wave beat it broadside, and those in the boat who were not thrown out had to jump into the vortex of fresh water entering the sea at Grange Chine. There were 14 of the ship's crew and 13 of the life-boat. There was considerable difficulty on account of the want of shore room, the sea reaching up to the cliff. There were plenty of willing helpers among the crowd, and at last the boat was safely housed, and the rescued and rescuers went home for a few hours' rest. The shipwrecked crew were housed in the village.

A few days later the ship broke up, and cases of 14 lb. tins of corned beef were washed up from Chale to Brighstone. Many were smashed, and loose tins were knocked into many shapes. Besides corned beef the vessel had a cargo of tinned prawns and tomatoes. Tomatoes were not so extensively grown as they are today, and the villagers did not know what they were until they were told. All the labels were washed off, so they had to open them to find out what the contents were. These unexpected supplies were a godsend to many a poor family.

By 1880 *Rescue* was proving to be too small and it was decided that she should be replaced. The new boat chosen was a boat purchased with funds raised by the staff and cadets of the Thames Nautical Training College's training ship *H.M.S. Worcester* and aptly named, *Worcester Cadet*.

Presentation of a Life-boat by the Cadets of the Thames Nautical College Ship Worcester
"Three Cheers for the Life-boat"

with acknowledgment to the Illustrated London News Picture Library

WORCESTER CADET

COMMISSIONING

From accounts in the Illustrated London News and in the Marine Engineer 1st May 1880.

The ceremony of presenting the *Worcester Cadet* life-boat to the representatives of the Royal National Life-boat Institution took place on the River Thames at Greenhithe on Tuesday 20th April 1880. The life-boat being brought alongside the training ship *H.M.S. Worcester* and hailed with three "English sailor like" cheers by the Worcester cadets.

The Management Committee of the *Worcester* were represented by the chairman Mr. Chambers and by Captains Carr and Maitland, the Royal National Life-boat Institution were represented by Admiral Ward and Captain St. Croix and amongst the company were Captain Ward, the donor of the largest subscription, and Captain Boxer of the *Chichester*.

The proceedings opened with a prayer for the success of the boat in her work of mercy offered by the Rev. Mr. Gilmore, following this Mr. Chambers invited the Captain-Superintendent of *H.M.S. Worcester,* Commander J. Henderson-Smith, R.N.R., to present the boat.

Commander Henderson-Smith was immensely cheered on rising to make the presentation and his speech was frequently interrupted to bursts of cheering from the boys. The first large donor he said was Jack Robinson, a well known former cadet and though he was bound to acknowledge the munificent subscription of Captain L. R. Ward amounting to £53, the fund of £700 had been largely made up of smaller subscriptions from the boys and their friends.

In accepting the new boat Admiral Ward, after thanking all concerned, described the virtues of the new boat and her self-righting features. An adjournment was then made to the side of the ship, where Mrs. Smith completed the ceremony by dashing a bottle of wine against the boat's bows amid the cheering of the boys.

The boats capabilities for righting herself after being capsized were next exhibited two or three times, affording much pleasure, mingled with wonder, to the enthusiastic cadets. The cadets were then divided into batches and they stepped into the life-boat, and were rowed around the ship.

ARRIVAL AT BRIGHSTONE

Following her commissioning and subsequent testing in London, *Worcester Cadet* arrived at Cowes on the 10th August 1880 and it was decided that in order to afford the inhabitants of the island an opportunity of seeing the new life-boat she should travel overland to Brighstone. Accordingly the next day she set off on her transporting trolley, with crew equipped ready for launching and drawn by six horses. At Newport she was met by the band of the headquarter company of the Isle of Wight Rifle Volunteers who escorted her to the Guildhall, where Captain the Hon. H. W. Chetwynd, R.N. (Inspector of Life-boats) presented the boat. The Rev. Canon Connor, the Rural Dean, received it on behalf of the Isle of Wight Life-boat Board and addressed the crew. The Mayor, Herbert J. Orchard, Esq., J.P., followed with an address of welcome, and at the conclusion the band played the National Anthem, the crew standing with oars erect, after which the band led off striking up "Hearts of Oak" and accompanying the boat as far as Carisbrooke Vicarage.

Worcester Cadet *at Newport Guildhall 11th August 1880*

The Brighstone Life-boat crew of 1881. Second Coxswain T. Cotton and Coxswain M. Munt, drowned in the Sirenia *rescue, are on the left. Bowman R. Buckett extreme right. Below centre left, Coastguard Officer S. Tubbs, Canon W. F. Heygate, Rector of Brighstone and honorary secretary of the Life-boat Committee, Mr. George Wyatt of Newport (honorary treasurer of the Isle of Wight Life-boat Board) and James Buckett, first coxswain, who in early life was sent into the Royal Navy for five years as a punishment for smuggling but was discharged three years later with a good conduct record*

That evening *Worcester Cadet* arrived at Brighstone and was parked opposite Wayte Court pond. Next morning she was taken across the meadows, the lane being too narrow for her, through Marsh Green to the boathouse at Grange Chine where a short service was taken by the Rector of Brighstone, the Rev. W. E. Heygate, ending with the sailor's hymn "Eternal Father, Strong to Save". Speeches of welcome then followed and the boat was launched and exercised. In the evening the crew had a special supper in the School Room. The Parish Record tells that so earnest were the crew to do right that they resolved of themselves that if any man got tipsy between Cowes and Newport in fetching the boat he should forfeit his day's pay. This fine was happily not afflicted on anyone.

So *Worcester Cadet* passed into service.

BRIGHSTONE GRANGE STATION

ROYAL NATIONAL LIFE-BOAT INSTITUTION

The life-boat *Worcester Cadet* was placed on this station by the R.N.L.I. in 1880. The cost of the boat was defrayed from the *Worcester Cadet* life-boat fund which was raised by the exertions of Captain J. H. Smith late Captain-Superintendent of the training ship *Worcester* stationed at Greenhithe, Kent.

The following is a list of its services:—

1881 January	27th	Steamer *Claremont* of Newcastle	Lives saved 6
1884 January	3rd	Steamer *Duke of Westminster* of London	Lives saved 21
1884 January	5th	Steamer *Duke of Westminster* of London	Lives saved 52
1888 March	9th	Ship *Sirenia* of Glasgow	Lives saved 6
1888 March	10th	Ship *Sirenia* of Glasgow	Lives saved 24
1890 April	6th	Schooner *Dizzie Dunlop* of Porthmadog	Lives saved 2
1890 April	7th	Schooner *Caboceer* of Chester	Lives saved 4
1892 February	1st	Steamer *Eider* of Bremen	Saved mails & 88
1892 February	2nd	Steamer *Eider* of Bremen	Saved specie & 41

In 1892 The *Worcester Cadet* was replaced by the life-boat *Joe Jarman*.

By order
Charles Dibbin
Secretary

Royal National Life-boat Institution 1892

The Rescue Board now exhibited in Brighstone Village Reading Room

THE RESCUES

27th January 1881 — S.S. CLAREMONT

The steamship *Claremont* of Newcastle, Captain Bainbridge, master, 668 tons register, from Vallerigan to Middlesbrough loaded with iron and copper ore, came ashore in dense fog below Whale Chine. When she struck the crew started her siren blowing and lashed down the tongue so that the siren kept blowing until the steam ran out. Hearing the siren the *Worcester Cadet* was launched, meanwhile the crew of the *Claremont* had taken to the boats. Heading towards the wreck the life-boat came up with one of the ship's boats containing the mate and five members of the crew whom she rescued. The other boat containing the master and the remaining 12 crew members was picked up next morning by the tug *Victoria* of Cowes, which had come out to attempt to salvage the steamer. The *Claremont* became a total wreck and portions of her hull strewed the beach for many years.

3rd January 1884 S.S. DUKE OF WESTMINSTER

The steamship *Duke of Westminster,* 2,726 tons register, from Brisbane to London with passengers and general cargo, ran ashore on Atherfield Ledge at seven o'clock in the evening in dense fog. With their usual promptness the crew of the *Worcester Cadet* mustered, launched the boat and proceeded to her assistance, where she took off 20 passengers and landed them at Brighstone.

Attempts were made to get the ship afloat with kedge anchors and she was pulled off, but as her propellor was broken she was helpless and soon ran fast aground again. Tugs arrived and attempted to haul her off but the hawsers parted and the attempt ended in failure. Some 70 Indian crew members were taken off and sent to the Asiatic Sailors' Home and shore hands were taken aboard to start getting out her cargo, the shore becoming strewn with oranges and coconuts much to the delight of young and old.

During the work of unloading the weather deteriorated and the life-boat had to be sent for and landed 50 of the men. Eventually with the aid of two tugs from Liverpool she was towed off and taken to Southampton for repairs. After which for many years the "Duke", as she was affectionately known, could be seen passing up and down the Channel on her lawful occasions.

9th March 1888 S.V. SIRENIA

From an account by "Vectensis" with acknowledgment to the Isle of Wight County Press.

The *Sirenia,* a fine well-found sailing ship of some 1,200 tons register was on a voyage from San Francisco to Dunkirk with a cargo of wheat when she struck the Atherfield Ledge in thick fog. There was an exceptionally heavy sea running – the aftermath of a series of gales. Her master, Captain McIntyre, had his wife and three children with him and a crew of 26.

She ran on the rocks at about three o'clock in the afternoon of Friday 9th March 1888. The *Worcester Cadet* was immediately launched. After a rough trip she reached the stranded ship and took off the Captain's wife and children, a female servant and an apprentice and landed them on Atherfield beach. Captain McIntyre, thinking that his ship was in no immediate danger, chose to remain on board with his crew for the time being, but arranged with Coxswain Munt that if necessary the life-boat should return at low water, just after midnight, when it was hoped that the sea would not be so rough.

In accordance with the arrangement, and knowing the danger of the ship breaking up in the heavy seas, Coxswain Munt and his crew again put out to the rescue just after midnight. The carriage, without which the life-boat could not be relaunched, had been brought from Brighstone and lowered over the cliff.

By this time conditions had become even worse, but nothing daunted the

The Sirenia *ashore on Atherfield Ledge*

The graves of Moses Munt and Leonard Dozier in the churchyard of St. Mary's, Brighstone

"A BRIGHSTONE BOY."

Among the many wreaths placed by kind and sympathising hands on the coffin of the American, Leonard Dozier, was one chiefly composed of wild-flowers, arranged round a child's hoop (apparently), and ticketed "From a Brighstone Boy."

"When saw we thee a stranger and took thee in? . . . . And the King shall answer . . Inasmuch as ye have done it unto one of the least of these My brethren, ye have done it unto Me."

I gazed upon the handiwork
  Of one who could employ
His heart and hands in sympathy,
  And sign "A Brighstone Boy."

And as I mused, I thought, perchance
  It was his only toy
That lay concealed beneath the flowers
  Plucked by "A Brighstone Boy."

I sought to find, but found him not—
  For O! 'twould give me joy
To be allowed to grasp the hand
  That wrote "A Brighstone Boy."

No! none could tell from whence it came,
  No *show* did he employ;
His work lies there—but he remains
  To me—"A Brighstone Boy."

Bright be thy path and glad thy lot,
  True bliss without alloy;
For thou hast "ta'en the stranger in,"
  Though but "A Brighstone Boy."

And when on earth thy course has run,
  May willing hearts employ
Their hands to do, as thou hast done,
  For thee "A Brighstone Boy."

And when before the Judgment Seat
  May angels shout for joy,
And bid thee "Welcome," when He says—
  'Well done" thou "Brighstone Boy!"

Brighstone,
  March 14, 1888. F.W.C.

life-boatmen pulled their craft through the breakers and safely reached the *Sirenia* and took off 13 members of her crew, but then came tragedy. As the *Sirenia* was stem-on to the beach, in line with the wind, there was no lee shelter for the life-boat. At the first attempt to make fast to the ship a three-inch hawser snapped, but a stouter line was passed from the ship to the life-boat and this held while the 13 men jumped aboard as the life-boat rose on the huge seas.

It was Bowman Robert Buckett's job to cut the hawser when the coxswain chose what he considered to be a favourable moment. but as he obeyed the order a huge sea struck the bow. With the weight of the rescued men in the stern the effect was to raise the life-boat to an almost perpendicular position and the sea carried her towards the shore at a frightening pace. Even so, all might still have been well had not a cross-sea, breaking over the Ledge, caught the life-boat on the starboard side and capsized her. Some of the life-boatmen, remembering their drill, gripped the thwarts as she rolled over, but the coxswain and second

74 PUNCH, OR THE LONDON CHARIVARI. [FEBRUARY 13, 1892.

MR. PUNCH TO THE LIFEBOAT-MEN.

Cartoon by Linley Sambourne
With acknowledgment to the proprietors of Punch

coxswain, in the stern, as well as the sailors from the *Sirenia,* had no chance to do this and when the life-boat righted herself Moses Munt, Tom Cotton and two of the rescued men were missing. It was marvellous in such a sea that all but these four of the 26 men in the life-boat managed to reach her again.

Perished with cold and almost on the point of collapse, the rest of the crew pulled into the beach where both rescuers and rescued were given succour by those who had watched the capsizing with anxious fears.

When the life-boat grounded one of the crew Frank Downer and one of the rescued sailors Leonard Dozier, a young American Seaman, were entangled in ropes hanging over the bow. Downer soon recovered but Dozier was dead when he was released. The body of the coxswain came ashore near the boat soon after the tragedy and that of the second coxswain was found near Blackgang Chine two days later.

In spite of the disaster a few hours previously, and in the face of a raging sea, the *Worcester Cadet* once more went afloat with Rufus Cotton as coxswain and a crew composed of three men of her regular crew and volunteers from Brighstone, Blackgang, Ventnor, Niton, Sandown and as far away as Bembridge.

One of the three members of the Brighstone crew who went out for the third time was a mere youth of 16, Frank Salter, who said "I've been out twice, so I can go again". The others were Rufus and David Cotton. They were awarded the silver medal of the R.N.L.I.

The Brighstone crew when the life-boat capsized were:— Coxswain Moses Munt, Second Coxswain Thomas Cotton, Bowman Robert Buckett, James William (Rufus) and David Cotton, Albert Whitewood, Frank Buckett, Robert Frank and Joseph Salter and George and Frank Downer.

Those who volunteered to join the crew on the third trip were:—George Attrill (Bembridge), John Cotton (Brighstone), Charles Kemp (Sandown), Percy Wheeler (Blackgang), Arthur Sothcott and William Buckett (Sandown) and Fred Bastiani and Charles Orchard (Niton).

Attempts to refloat the *Sirenia* failed and she was rapidly battered to pieces.

The funeral of the three victims took place the following Wednesday at the church of St. Mary, Brighstone. The service was taken by the Rev. G. E. Jeans, Vicar of Shorwell, and Rev. F. B. Lipscomb, Curate of Brighstone, the illness of the rector, Rev. W. F. Heygate, preventing his attendance. The coffins of the two life-boatmen were borne by crews of the Brighstone and Brook life-boats and that of the American sailor by four coastguardmen. The graves of Coxswain Munt and Seaman Dozier lie side by side in the churchyard a few yards south of the church tower and some yards away lie the remains of Thomas Cotton, the second coxswain.

6th January 1890 — Schooner DIZZIE DUNLOP
7th January 1890 — Schooner CABOCEER

The schooner *Dizzie Dunlop*, from Port Madoc to Kiel, with a cargo of slates came ashore on the Atherfield Ledge on the evening of Sunday 6th January 1890. The crew with the exception of the master and the mate, who had refused to leave, were taken off by the coastguards. Eventually the life-boat was called out, the *Worcester Cadet* was launched and going alongside the stricken shooner took off the master and the mate who were by this time in a very exhausted condition.

S.S. Eider *ashore on Atherfield Ledge from a contemporary painting*

Atherfield, Brook and Brighstone life-boats drawn up on Atherfield beach

On their return to shore the crew were informed that two further vessels were on the 'Ship' ledge near Brighstone, the life-boat was immediately relaunched and the two crews were brought safely ashore. The first of these was the schooner *Caboceer* of Chester from Swansea with a cargo of coal, the second was the ketch *Cameo* bound from Penzance to Woolwich Arsenal with a cargo of cracked granite. Whilst alongside the *Caboceer* a rocket fired from the shore passed through the side of the vessel, narrowly missing the captain's head.

The steam tug *Malta* came out from Cowes and stripped the *Caboceer* and the cargo from the *Dizzie Dunlop* was salvaged by the Atherfield fishermen under a contract of 50 per cent of the value. The *Cameo* was eventually towed to safety.

The crews of all three ships were provided for and sent to their homes free by the agent of the Shipwrecked Mariner's Society. The Brighstone life-boat crew were, said the Isle of Wight County Press, entitled to high praise for their exertions on this occasion.

31st January 1892 S.S. EIDER

The North German Lloyd Company's Atlantic steamer *Eider*, bound from New York to Bremen via Southampton, struck on Atherfield Ledge in thick fog just before ten o'clock on the evening of Sunday 31st January 1892.

The *Eider,* an iron and steel single screw steamer of 4,719 tons register, built by Fairfield & Co., on the Clyde in 1883, was on a regular passage with a crew of 163, 227 passengers, 500 sacks of mail, £300,000 worth of specie consisting of boxes of gold and silver ingots and a cargo of baled cotton and maize.

Under the command of Captain Heineke, one of the most experienced and trusted captains in the company's service, she had left New York on the previous Saturday passing Sandy Hook at 11 o'clock and had made a rapid passage averaging 16 knots.

Due to the thick fog they had missed sighting the Needles lighthouse and the entrance to the Solent, running on under easy steam taking soundings as they went, she never the less ran onto the Atherfield Ledge at high tide half a mile from the beach. As she struck the fog lifted and the watchers on the shore saw alternatively her green and red side lights as she twisted around four or five times before settling in a cradle of rocks immediately opposite the Atherfield Coastguard station.

The vessel had not long struck before news of the disaster had spread and the Atherfield life-boat, the *Catherine Swift,* was launched and got alongside, but Captain Heineke, entertaining strong hopes that she could be pulled clear, declined assistance. With the weather worsening the life-boat returned to the *Eider* in the early hours of the next morning but still her services were declined and she returned to the shore bringing only telegrams. Tugs had by now arrived and attempts were made to pull her off but she remained fast.

As daylight broke, with the weather still bad, the Brighstone and Brook life-boats were summoned as well and after a hard pull the *Worcester Cadet* was the first alongside and took off the first batch of passengers, landing them safely at 11 o'clock. During this time the crowds on the cliffs had been watching the Brook life-boat, the *William Slaney Lewis,* toiling up channel and after five hours' hard rowing were alongside to take off more of the women and children passengers. Meanwhile the Atherfield boat had, with great difficulty, been relaunched and got alongside. So the three boats worked all day bringing ashore 13 or 14 passengers at a time. By the Monday evening most of the passengers were ashore and on the Tuesday, despite a very heavy sea running all day, the crew were got ashore. By nightfall all on board had been safely rescued. Most of the passengers were in an exhausted condition from fright and exposure and as soon as they were lifted out of the boats they were taken to the homes of the villagers, to Brook House the home of the squire Mr. Seely, and as far afield as the Bugle Hotel in Newport.

On the Monday *Worcester Cadet* had rescued 88 and on the Tuesday was to rescue another 41. As well as passengers and crew she is credited with bringing ashore much of the mail that the *Eider* was carrying.

On the Wednesday, permission having been obtained from the headquarters of the R.N.L.I., the life-boat crews were employed in bringing the specie and the remainder of the mail ashore. The men were to be paid £1 for every ingot of silver and £2 for each case of gold. As the specie was landed it was taken in charge by the coastguards armed with swords and loaded revolvers. It was taken up the cliff and placed in wagons for conveyance to Cowes, armed men being placed on each wagon, the procession of wagons attracting much attention in Newport and Cowes.

Among the visitors to the scene of the disaster were H.R.H. Prince Henry of Battenberg, the Governor of the Isle of Wight, and on the Friday following the wrecking of the *Eider* the Prince of Wales accompanied by Prince George (later King George V) drove to Atherfield and was introduced to the coxswains and bringing the command of Queen Victoria "that her warm appreciation of the gallant conduct displayed by the crews of the Atherfield, Brighstone and Brook life-boats in saving crew and passengers of the *Eider* be conveyed to them".

The *Eider* was eventually refloated on 29th March and towed much damaged to Southampton.

The coxswain of the *Worcester Cadet,* James H. Cotton, received the silver medal of the R.N.L.I., their thanks on vellum and, as did each of the other boat's coxswains, from his Imperial Majesty the Emperor of Germany, a gold watch bearing the Emperor's portrait and initials inscribed:—

> "We, William by the grace of God, Emperor of Germany and King of Prussia, grant to the coxswain of the life-boat *Worcester Cadet* of Brighstone Mr. James Cotton, this acknowledgement of help rendered to the crew and passengers of the stranded Mail steam-ship *Eider* of Bremen, when in distress at sea".

Amongst the many letters the coxswain received was one from Captain Henderson-Smith expressing the congratulations of himself and the cadets of *H.M.S. Worcester*, indicating how the original donors had, over the years, maintained their interest in the boat which they had presented to the R.N.L.I.

So ended in triumph *Worcester Cadet's* career as a life-boat. Later in 1892 she was withdrawn from service and replaced by the new life-boat *Joe Jarman.*

THE STRANDING OF THE *EIDER*

THE *Eider* rode in the open sea
With her safety in God's own hand
For a thousand miles—ay, two, and three,
With never a sight of land.

A shell of steel on the world of waves
That severs the hemispheres,
That covers the depths of a thousand graves
And the wrecks of a hundred years.

She bore, unhurt, through the storm-god's din,
Through shower and shade and sheen,
With the death without and her lives within,
And her inch of steel between.

From the port behind to the port beyond,
With never a help or guide,
Save the needle's point and the chart he conned,
The master has fought the tide.

On the bridge, in the Sunday twilight dim,
He has taken his watchful stand;
And he hears the sound of a German hymn,
And the boom of a brazen band.

He looks for the lights of the royal isle,
Ahead, to left and to right;
Below there is music and mirthful smile,
For land must be soon in sight.

In sight? Not yet! for a fog creeps round
And the night is doubly dark.
"Slow speed! Hush! is it the fog-bell's sound,
Or the shriek of the siren? Hark!"

The fog-bell clangs from its seaward tower,
And the siren shrills in fear;
But the vapours thicken from hour to hour,
And the master cannot hear!

On the seaward headland, the beacon's blaze
Like a midday sun would seem,
But its warning rays are lost in the haze,
And the master sees no gleam!

"How goes the line? There is time to save!"
"It is ten fathom deep by the log."
"We have not tarried for wind or wave,
We cannot wait for the fog."

On, on! through the dark of a double night;
On, on—to the lurking rock!
No sound, no gleam of a saving light
Till the *Eider* leaps to the shock.

All night she bides where the sea death hides,
And her passengers crowd her deck;
While the leaping tides laugh over her sides
And sink from the stranded wreck.

The *Eider* has gold, she has human lives;
But these can assist no more.
Pray, pray, ye German children and wives,
For help from the English shore!

A signal is sent, and a signal is seen,
And a life-boat—ay, two and three,
From the shore to the vessel their crews row between,
And fight with the stormy sea.

They fight day and night, as true-Englishmen can,
With the servants they rule the waves;
And the *Eider's* four hundred are saved to a man
From the terror of sea-bed graves.

The *Eider* bides, all broken and bent;
With the tide she shivers and starts,
And stands—for a time—as a monument
Of the courage of English hearts.

But longer lasting, the memoried grace
Of a noble deed and grand
Will knit the hearts of the English race
To the hearts of the Fatherland!

Anonymous.

"ON THE BLAZON'D SCROLL OF FAME."

[To each man of the Crews of the three Life-boats stationed in the Isle of Wight, at Brighstone, Brook and Atherfield, respectively, *Mr. Punch* has had pleasure and pride in presenting an illuminated copy of the Picture and Poem entitled "MR. PUNCH TO THE LIFE-BOAT MEN," which appeared in his issue of February 13. The names of the coxswains and crews of these three boats, the *Worcester Cadet*, the *William Slaney Lewis*, and the *Catherine Swift*, are inscribed thereon (as they should be in the memories of all true Britons), as follows:—Of the *Worcester Cadet*, JAMES COTTON (Coxswain), ROBERT BUCKETT (Second Coxswain), ROBERT SALTER, WILLIAM BARTON, FRANK EDMUNDS, FRANK BUCKETT, GEORGE NEW, GEORGE MORRIS, GEORGE SHOTTER, GEORGE HAWKER, EDGAR WHITE, WILLIAM MERWOOD, and JAMES HEDGECOCK.

Of the *William Slaney Lewis*, JOHN HAYTER (Coxswain), BEN JACOBS (Second Coxswain), ROBERT COOPER, W. JACOBS, J. COOKE, G. WHITE, W. CASSELL, T. HOOKEY, J. NEWBURY, J. COOPER, J. HOOKEY, R. WOODFORD, M. CASSELL, WILLIAM HAYTER, W. BLAKE, and W. HOOKEY.

Of the *Catherine Swift*, WILLIAM COTTON (Coxswain), DAVID COTTON (Second Coxswain), JAMES COTTON, THOMAS COTTON, FRANK COTTON, JOHN COTTON, CHARLES COTTON, WALTER WOODFORD, WALTER WHITE, CHARLES HARDING, and B. WHILLIER.

These names thus receive—as they deserve—honourable record "For distinguished bravery and gallant conduct whilst on duty on the occasion of the wreck of the s.s. *Eider*, January 31, 1892."]

ON the Scroll! And why not? Be you sure that it bears
Many entries less worthy of record than theirs,
The rough sea-faring fellows, whose names now go down,
With applause from their Sovereign to swell their renown,
To posterity's ears. And right pleasantly, too,
They should sound on those ears; for, run over each crew
And you'll find that those names have a true homely smack
Both of country and kinship; there's JIM, there is JACK,
There is BOB, there is BILL, TOM and GEORGE, CHARLIE, FRANK;
Can you not hear them sound o'er the waves as in rank
They go down to their work, ringing right cheery hail
Through the shrieks of the storm that shall not make *them* pale,
Those bold Britons? They're brothers, sires, cousins, and sons,
For see how the "family name" through them runs
Those COTTONS could make up a crew at a pinch!
Whilst the HOOKEYS and WHITES from that task need not flinch.
Yes, these names sound as well on the Scroll, after all,
As NAPOLEON or CÆSAR; and when the Great Call
Of the last human Muster Roll comes, some plain "BILL,"
Whose business was rather to save than to kill,
May step before mad ALEXANDER.
Well, brothers,
(You BUCKETTS, and WOODFORDS and COOPERS and others,
Whose names he need hardly string into his rhymes,)
Punch hopes you may look on this Record sometimes
With pleasant reflections. Mere words, he well knows,
Will not—"butter your parsnips"—(to put sense in prose):
But you have his hearty good will, and you know it,—
Right gladly he takes this occasion to show it!
And when or wherever *another* should come,
Be sure your friend *Punch* won't be careless or dumb!

YOUR hand, lad! 'Tis wet with the brine,
and the salt spray has sodden your hair,
And the face of you glisteneth pale with the
stress of the struggle out there;
But the savour of salt is as sweet to the sense
of a Briton, sometimes,
As the fragrance of wet mignonette, or the
scent of the bee-haunted limes.

Ay, sweeter is manhood, though rough, than
the smoothest effeminate charms
To the old sea-king strain in our blood in the
season of shocks and alarms,
The soft-hearted shore-going critics of conduct
themselves would not dare,
The trivial cocksure belittlers of dangers they
have not to share,
Claim much—oh *so* much, from rough manhood,—unflinching
cool daring in fray,
And selflessness utter, from toilers with little
of praise, and less pay.

Her heroes to get "on the cheap" from the
rough rank and file of her sons
Has been England's good fortune so long, that
the scribblers' swift tongue-babble runs
When the winds and the waves and the rocks
make a chaos of danger and strife;
And the need of the moment is pluck, and
the guerdon of valour is life.

That guerdon you've snatched from the teeth
of the thundering tiger-maw'd waves,
And the valour that smites is as naught, after
all, to the valour that saves.
They are safe on the shore, who had sunk in
the whirl of the floods but for *you!*
And some said you had lost your old grit and
devotion! We knew 'twas not true.

To the old easy tune without thought. "Gallant
sea-dogs and life-savers!" Yes!
But poor driblets of lyrical praise should not
be their sole guerdon, I guess.

On the coast, in the mine, at the fire, in the
dark city byeways at night,
They are ready the waves, or the flames, or
the bludgeoning burglar to fight.
And are *we* quite as ready to mark, or to
fashion a fitting reward
For the coarsely-clad commonplace men who
our life and our property guard?

A question *Punch* puts to the Public, and on
your behalf, my brave lad,
And that of your labouring like. To accept
your stout help we are glad:
If supply of cheap heroes *should* slacken,
and life-saving valour grow *dear*—
Say as courts, party-statesmen, or churches
—'twould make some exchequers look
queer.

Do we quite do our part, we shore-goers?
Those lights could not flash through the
fog,
And how often must rescuer willing lie idle
on land like a log
For lack of the warning of coast-wires from
lighthouse or lightship? 'Tis flat
That we, lad, have not done *our* duty, until
we have altered all that.

Well, you have done yours, and successfully,
this time at least, and at night.
All rescued How gladly the last must have
looked on that brave "Comet Light,"
As you put from the wave-battered wreck.
Cold, surf-buffeted, weary, and drenched,
Your pluck, like the glare from that beacon,
flamed on through the dark hours unquenched.

Nor then was your labour at end. There was
treasure to save and to land.
Well done, life-boat heroes, once more!
Punch is proud to take grip of your hand!
Your QUEEN, ever quick to praise manhood,
has spoken in words you will hail,
And 'twere shame to the People of England,
if they in their part were to fail.

With acknowledgment to the proprietors of Punch

Life-boat Joe Jarman *on boat carriage at the naming ceremony on 17th October 1892 Crew and number of times afloat on life-boat service, James Cotton (coxswain) (35). Robert Buckett (second coxswain) (33), Frank Symonds (22), Frank Buckett (18), George Salter (15), William Merwood (15), William Holbrook (2), Walter Cotton (2), Victor Cotton (2), Fred Cotton (1), Robert Buckett (1), George Salter (29), Charles Holbrook (0)*

THE 'JOE JARMAN'
and
THE 'BROTHERS BRICKWOOD'

JOE JARMAN

The new boat *Joe Jarman* was purchased from a legacy of a Miss Jarman of Slough. She arrived on station on 17th October 1892 having been brought round by steam launch from Portsmouth. The two boats were exchanged at sea and *Joe Jarman* was brought ashore. A fortnight later on November 3rd a benediction service was held at the boat-house and the new boat named by Mrs. Seely.

Joe Jarman was to remain in service for 15 years but her name does not feature on any of the rescue boards that survive. The reason for this was that with the coming of steam the rate of maritime accidents fell dramatically and the services of the life-boats everywhere were called upon less frequently. Despite the non-appearance of her name on the rescue boards she was not inactive.

On 17th January 1894 she was called out early in the morning to assist a steamer ashore at Brook Point but when she arrived, having been taken along the coast road to be launched, it was found that the survivors had already been taken off by the Brook boat and the laconic statement "No Service" was recorded in the minute book. Again three months later on 13th March she stood by a dismasted sailing vessel in distress some five miles off the boat-house, but a steamer hove into sight and took the hapless vessel in tow and "No Service" was once again recorded.

On 9th March 1893 she stood by the steamer *Princess Henrietta* out of Ostend and on 17th January 1894 she went to assist the the S.S. *Ossian of Leith* ashore at Brook only to find once again that the Brook boat had beaten her to the wreck. Numerous other entries in the minute book show her assisting unknown vessels in distress, attending others only to find the crews taken off by tugs and other ships assisting. On another occasion on 19th February 1897 she stood by the wreck of a full rigged ship aground at Atherfield when this time it was the Atherfield boat, whose coxswain was another of the Cotton family one William, nicknamed Rufus, who carried out the rescue whilst *Joe Jarman* stood by.

Despite her lack of lives saved she was obviously kept busy during her 15 years at Brighstone Grange and all the time the drills and practices continued. A sad entry in the minute book concerns one such practice when old Ben Cooper fell under the front wheel of the carriage and died the next day from injuries and shock. At the end of her service at Brighstone she was returned to London and converted for demonstration purposes. After further service at Cork in Southern Ireland she was sold in 1918 for £40.

THE BROTHERS BRICKWOOD

The *Joe Jarman* had been launched at 5 a.m. on the morning of 30th May 1907 and the crew took her rowing and sailing to Lymington and the new life-boat *The Brothers Brickwood,* built with funds provided in a legacy from a Miss G. D. Brickwood of Boscombe, was brought back to Brighstone Grange that evening. She was six inches broader in the beam than her predecessor and this had necessitated the construction of a new carriage. The minute book records that the old carriage, which had served the station so faithfully, was taken by the Shepard brothers to Newport pulled by four horses under the control of two carters, the procession being led by a small boy of nine waving a red flag. Several gate posts had to be taken up at Marsh Green and Waytes Court to let the carriage through. The procession returned immediately with the new enlarged carriage.

The naming ceremony in the afternoon of 30th May 1907 was a great occasion. The County Press records that the inhabitants of Brighstone had turned out en masse and the names recorded of those present included many dignitaries connected with the R.N.L.I. and the coxswains of all six of the Island life-boats were present. *The Brothers Brickwood* lay in her carriage bedecked with flags and an anchor composed of wild flowers hung over her bow. The ceremony started with a short service of dedication led by the Rev. E. W. Silver, the Rector of Brighstone and Rev. G. E. Jeans, the Vicar of Shorwell and the surpliced choir sang the sailors' hymn "Eternal Father, Strong to Save". After an introduction from Sir Charles Seely, Bart, the President of the Isle of Wight Life-boat Board, Captain Keppel Foote, the local Inspector of Life-boats, formally handed the boat into the safe-keeping of the local committee. The naming ceremony was carried out by Sir Charles Seely's daughter who at the second attempt broke the

Brothers Brickwood *ready for sea trials showing tipping plates in position*

Crew at this time James Cotton (coxswain), Robert Buckett (second coxswain), Frank Edmunds (bowman), Frank Buckett, George Shotter, William Holbrook, Walter Cotton, Robert Buckett, jnr., Charles Holbrook, John Creeth, Edwin Munt and Aaron Downer

traditional bottle of champagne over the bows. The crew dressed in their smart new jerseys emblazoned "RNLBI Brighstone Grange", a gift of the local committee, launched the boat and accompanied by Captain Foote carried out a sailing and rowing exercise.

The new boat was to prove popular with her crew who reported that she was more "sea-kindly" in a rough sea and being lighter that her predecessor was easier to manhandle when launching and recovering. She was to be involved in one incident where her name appears on the rescue board.

20th October 1909 — Barge WILLIAM AND MARY

On Wednesday 20th October 1909, at about six o'clock in the evening, the ketch-rigged barge *William and Mary,* captain and owner J. C. Lawrence. 115 tons register, of Maldon, bound from Calstock, near Plymouth, to Portsmouth and Chatham Dockyards and London, with a cargo of firebricks and tiles, was driven ashore about a quarter of a mile to the westward of Grange Chine.

She had left Plymouth on the Tuesday at 12 o'clock, and coming up Channel had had to face very bad weather. When off Portland the force of the gale snapped the sprit, and to steady the vessel they cut away the gear. With only the jib and foresail they struggled on against mountainous seas and tried to make the Needles, which, however, in the misty rain they failed to do, finding themselves, in a sudden clearing of the mist, close to the Island shore they tried their hardest to beat out into the Channel for the purpose of rounding St. Catherine's but it was impossible, and the ship was forced by the heavy seas and wind nearer and nearer the shore. Realising his danger, the captain endeavoured to locate a coastguard-station, off which he intended to drop anchor and signal for help. As darkness fell they began signalling with a flash-lamp, and getting an answering light from the shore they dropped the two anchors and waited.

The Brighstone coastguards had been watching the little vessel struggling against the heavy sea during the afternoon, and in response to the flashlight signal, the life-boat rocket and gun were fired. At 6.30, when Coxswain J. Cotton and his crew went to the rescue in *The Brothers Brickwood,* the barge had drifted to within a few hundred yards of the beach.

Before the life-boat got to them, however, one of the anchor cables had snapped and the vessel was dragging towards the shore. The captain's arm was in a sling, as he had sustained a broken wrist in endeavouring to hold the wheel against the heavy seas. The vessel was practically under water continually, and he was three times thrown over the wheel. The life-boat was soon alongside and got off Captain Lawrence and his mate Louis Hardy who, on landing, were at once taken to the life-boat house, where they were supplied with dry clothes and hot coffee. Before leaving the vessel the captain dropped two anchors, which held for some time, but as the tide rose the heavy seas broke the cables, and the ship was dashed right up to the edge of the cliff. The two men were accommodated in the village for the night, and both expressed their gratitude to the life-boatmen and the residents of the village, who had been so kind to them. The mate's dog, a foxterrier, remained on board the vessel throughout the night, but was safely rescued in the morning, much to the joy of its master.

Missing from his regular position on this occasion was the Second Coxswain Robert Buckett who had sustained a crushed finger a few days before at an inspection launch.

Like *Joe Jarman, The Brothers Brickwood* was kept relatively busy but again much less so than the *Rescues* and the *Worcester Cadet.* The minute book records occasions when she was launched and "No Service" was required and the last occasion recorded on 17th March 1914 recalls that the crew were mustered to assist the barquentine Balder in distress under tow. Although her sister boats at Atherfield and Brook were similarily alerted, no launch of any boat was required. So ended the life-boat service at Brighstone Grange.

After leaving Brighstone *The Brothers Brickwood* was put in reserve and later served at Scarborough from 1918 until 1924 when she was sold to the Latvian Life-saving Society for £150.

BROOK

Plans for life-boat stations at Brook and Brighstone were formed at the same time and both stations were opened on the 13th August 1860.

The life-boathouse at Brook was a low stone-built building constructed on open land given by Mr. Charles Seely at the top of Brook Chine facing the old coastal track from Freshwater to Chale, the present coast road. When first opened the life-boat was transported on a horse-drawn launching carriage similar to that in use at Brighstone, but towards the end of the life of the station a tractor was used to take the boat down the chine to the beach.

The first life-boat to be stationed at Brook was the *Dauntless* and she was to remain on station for seven years. Little is known of the number of times she was called out, her name is mentioned on the service board now hanging in Brook Church, but her record does not appear. That she was active is undoubted but the R.N.L.I. records show that no life-saving services were rendered.

In 1867 *Dauntless* was transferred to North Berwick and was replaced by a new boat, presented to the R.N.L.I. as a gift from the late Mrs. Marten of Upper Clapton and named the *George and Anne*.

The *George and Anne*, which remained in service for 17 years, served the station well and during her time 84 lives were saved from ships she had attended. In 1892 she was replaced by a new boat the *William Slaney Lewis* purchased through a legacy of Mr. W. S. Lewis of Wellington, Salop.

The new boat was very soon in action, only a few weeks after she had been delivered, in the ill fated and tragic attempted rescue of the passengers and crew of the *Sirenia*. She was to remain in service until 1907 when she was returned to London and sold to King & Co. in November of that year. During her time at Brook 171 lives were saved.

The last life-boat to be stationed at Brook was the *Susan Ashley* and the records show that in the 30 years she was to be in service, the longest service of any of the Brighstone Bay life-boats, she was launched relatively infrequently and her tally of lives saved was only 13. At the end of her service she was sold to a Mr. Hescroft of Poole who converted her into a yacht for £75. In September 1957 she was seen visiting Yarmouth, Isle of Wight as the auxiliary cruiser *Susan of Poole* and as late as 1977 was to appear in the film of Erskine Childers' famous thriller "The Riddle of the Sands".

Inextricably entwined in the story of the life-boats of Brighstone Bay is the name of the Seely family of Brook House. Mr. Charles Seely of Brook House was in the early days prominent as a benefactor donating land and funds towards the upkeep of the stations. He, and from 1887 his son Sir Charles Seely the first baronet, and their families were it seems ever present when a ship disaster struck, to organise sustenance and accommodation for the survivors and in arranging to have them transported to the mainland or even on occasion to their home port.

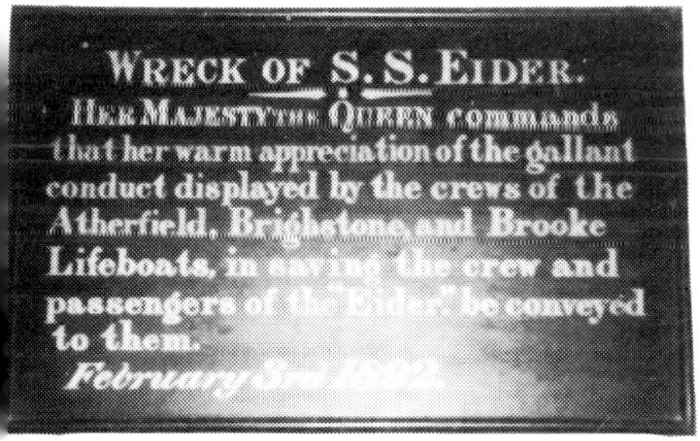

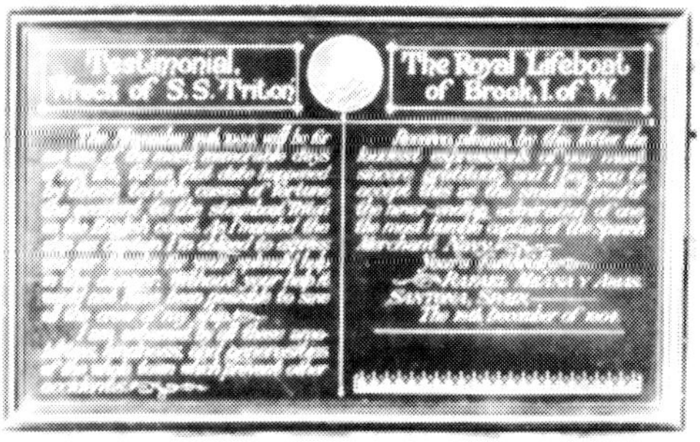

The life-boat boards in Brook Church

Jack Seely—later 1st Lord Mottistone
Photograph Lord Mottistone

The closest attachment of the Seely family, however, was to Brook through Sir Charles Seely's younger son John Edward Bernard, known as Jack, and later in 1933 to become the 1st Lord Mottistone. For 40 years Jack Seely was crew member and supporter of the station and during the 1930's served as a coxswain. Born on the 31st May 1868 he was educated at Harrow and Trinity College, Cambridge, he was called to the bar in 1897. He served in South Africa with the Imperial Yeomanry winning a Distinguished Service Order and was later to become Member of Parliament for the Abercromby Division of Liverpool and to hold high government office culminating as Secretary for War prior to 1914. Certainly the only Cabinet Minister to be a member of a life-boat crew, he played his full part when Parliamentary duties permitted him to be at Brook. His description in his book "Adventure" of a launching and the subsequent assistance given to a sailing ship in distress is a vivid example of the bravery of all the crew and exemplifies the friendship between the Lord of the Manor's son and the local fisherman and farmers who made up the crew and shore party. He modestly states in the same book that owing to frequent absence he had little to do with the splendid record of the Brook life-boat station. However, the name of J. E. B. Seely is forever held high in the annals of the Isle of Wight.

Other names closely associated with Brook life-boats were Tom Hookey, a particularly close friend of Jack Seely, his brother David the village blacksmith, also Ben Jacobs, for 37 years a member of the life-boat crew and coxswain for 25 of those years. Six foot tall and of immense strength, Ben Jacobs was a legendary hero in Brook, remembered for carrying his money in his hat and accordingly was rarely seen without it. His brother Phil Jacobs who also served in the crew with Roland Hayter who was the last coxswain during the First World War and up until the station closed.

15th November 1871 — Barque CASSANDRA

Early in the morning of Wednesday 15th November 1871, the barque *Cassandra* was wrecked in Compton Bay. The life boat *George and Anne* was called out immediately and by six o'clock in the morning was alongside the stricken vessel and had taken off half the crew. Four hours later having made a second trip she landed the remaining crew saving in all 21 lives. A few hours later the Government steam tugs *Camel* and *Grinder* arrived to give assistance with seamen and riggers on board.

19th January 1876 — Barque MIGNONETTE

Coming up the Channel on the morning of Wednesday 19th January 1876, the barque *Mignonette*, bound from Bernese to London with a cargo of sugar, rum and coconuts, made landfall at the Lizard at 4.30 a.m. after which the weather closed in and no further land was sighted.

During the following evening she ran aground on Brook Ledge. The *George and Anne* was launched and made slow progress crossing the heavy ground swell which was breaking over the reefs. Eventually she got alongside and after rescuing all 14 of the crew made her way back to safety.

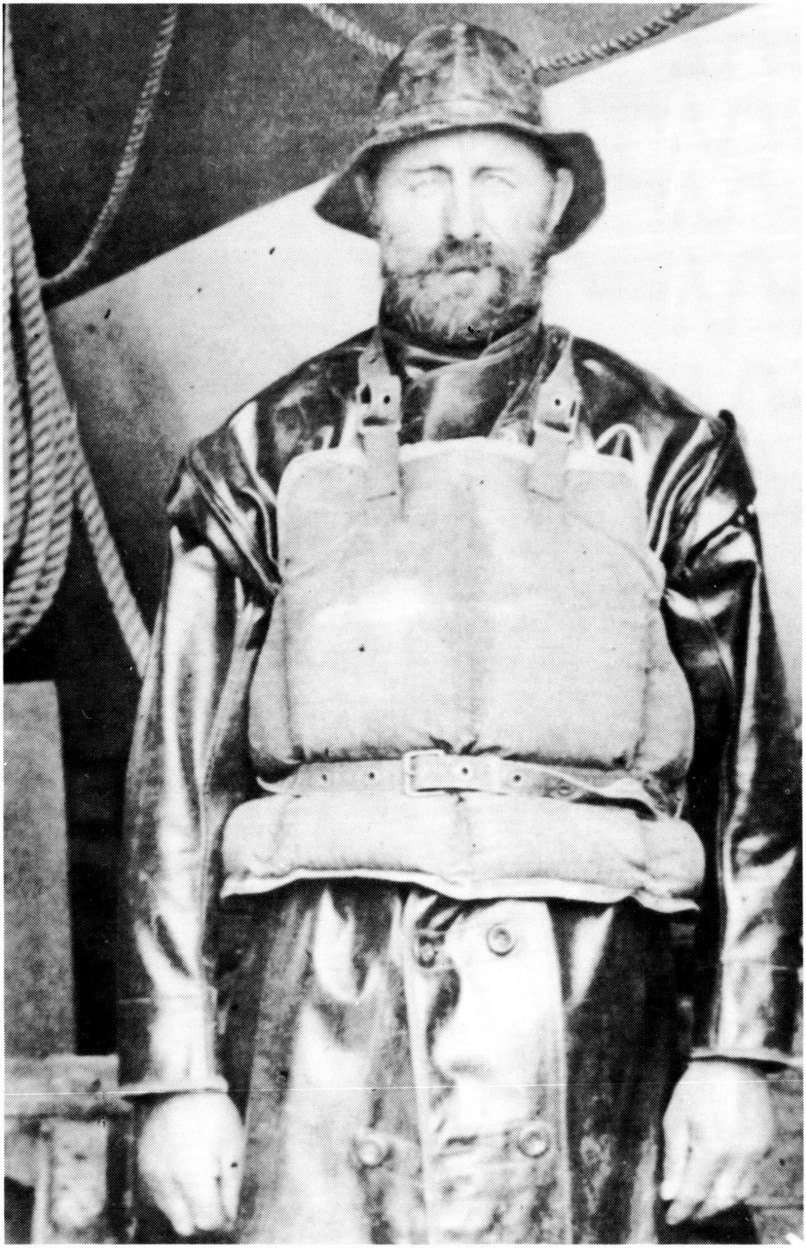

Ben Jacobs
Photograph Eric Toogood

9th March 1888 S.V. SIRENIA

The wreck of the *Sirenia* on Atherfield Ledge was to prove to be, from the life-boatmen's point of view, the most tragic accident to occur in the 57 years of the life-boat service in Brighstone Bay. Three crewmen were to lose their lives.

The full story of this wreck has already been told on page 14 but the men of Brook were to play an equal part with their colleagues of Brighstone.

Before the Brighstone life-boat was launched on her fatal trip from Atherfield, the *William Slaney Lewis* had also been sent to the rescue. The sea conditions inshore can be gauged from the fact that it was impossible to launch the life-boat at Brook and she was taken to Brighstone Grange. There an attempt failed, two members of the crew being injured, but not to be beaten, they took their boat to Chilton Chine and there successfully launched her.

Running before the gale, they had almost reached the *Sirenia* when a huge following sea was shipped, Second Coxswain Reuben Cooper and Ben and Phil Jacobs, members of the crew, were washed overboard. The two Jacobs held on to the lines and were hauled back into the boat, but Cooper disappeared. He was using the sweep oar at the stern and was probably flung some distance by the force of the sea. His cries for help were heard, both in the life-boat and on the *Sirenia,* but a search for him in the darkness was unavailing and his body was never recovered.

During the search, when the life-boat was near the *Sirenia,* there was a temporary lull in the gale, and Coxswain Hayter offered to take off the rest of the *Sirenia's* crew, but Captain McIntyre insisted that the search should continue and he and his crew would take the chance of rescue later. This the life-boatmen did and the boat was carried over the ledge into Chale Bay, too far to the eastward to row or sail back to the wreck. So they anchored until daybreak, when they gallantly attempted to pull back through a veritable hell of waters.

After spending the greater part of the night chilled to the bone, drenched to the skin, and without food or drink, the struggle to reach the *Sirenia* was too much for the Brook crew, and after a while they yielded to the shouted warnings of people ashore, set sail and beat back to their station, sadly mourning their lost comrade.

They were probably influenced to give up their exhausting struggle by the sight of the Brighstone life-boat being prepared for a third launch from Atherfield beach.

The Brook crew that night were:— Coxswain John Hayter, Second Coxswain Reuben Cooper, James Cooper, Ben and Phil Jacobs, John and Frank Cooke, John and James Newbery, Mark Cassell and Coastguards Saunders, Bridge and Ellender. Coxswain Hayter was honoured by the award of a silver clasp to his R.N.L.I. Silver Medal.

31st January 1892 S.S. EIDER

The wreck of the *Eider,* the story of which has been told on page 20 was a very public occasion. The crowds on the cliffs watched the distress of the ship ashore on Atherfield Ledge and the endeavours of the life-boatmen to save the passengers and crew. Amongst the scenes they saw enacted would have been the heroic five hour row by the men of Brook to bring the *William Slaney Lewis* from Brook Chine to Atherfield and no doubt the cheers rang out from the cliff top.

The men of Brook joined in fully with their colleagues from Brighstone and Atherfield. On the Monday they saved 90 lives and on the Tuesday a further 48. On the Wednesday they were kept busy ferrying mail and specie ashore. One young lady with her baby was later to make a formal complaint that she had got her feet wet!

Coxswain Hayter was to receive a third clasp to his R.N.L.I. Silver Medal and also the gold watch from the Emperor of Germany. Queen Victoria's message of congratulation is recorded on a painted board now hanging in Brook Church.

7th January 1894 S.S. OSSIAN

The steamship *Ossian* of Leith, 387 tons, owned by the London and Edinburgh Shipping Company, was on passage from Bordeaux to Leith with a cargo of claret, port wine, sheep skins, cork shavings and seed. Under the command of Captain W. Symon she left Bordeaux on the 4th January and had made a stormy passage across the Bay of Biscay. After passing Ushant early on the morning of 6th January she saw no land until running ashore on Brook Ledge at three o'clock on the morning of the seventh.

The *Ossian* had previously been sighted by the coastguard on duty, and by Ben Jacobs, the coxswain of the Brook life-boat. Seeing the need for prompt action he at once summoned his men by rockets and sent for the horses, so that by 3.40 the boat was afloat, and making an excellent journey through the surf, took off seven of the crew of the *Ossian*. The captain, with eight more men, decided to remain on board until the tide had fallen, when the life-boat was again summoned by signal, and all were safely landed by 11 o'clock. The Honorary Secretary and other members of the local committee were on the spot and had prepared tea and coffee, ready for the men on landing. A strong body of helpers rendered good service in launching and landing the boat. News of the disaster was wired at once to the owners of the vessel, to the R.N.L.I. and to others. The Brighstone Grange life-boat was also made ready, and, with her crew, arrived at the scene by the coastal track, but her services were not required.

This was the first occasion that Jack Seely rowed in the crew. Taking the place of a sick crew member he performed so well that Ben Jacobs, the coxswain, proposed him as a permanent member. He was duly enrolled and remained a member of the crew from that day onwards.

24th March 1895 S.V. NOORDSTER

Early on Sunday morning 24th March 1895 the Dutch brigantine *Noordster* (North Star), 183 tons, Captain Ufken, of Albasserdam, near Rotterdam, bound from Maracaybo, Central America, for Hamburg with a cargo of boxwood and divi-divi, came ashore off Brook Point.

The vessel, was sighted by the coastguards about 5.45 a.m., having evidently approached too close to land in the heavy fog which had prevailed over the Channel. On discovering their perilous position the crew at once proceeded to put their ship about to get into open water, but in doing so their bowsprit and head-sails were carried away by the wind and the ship became unmanageable, and before she could be repaired she struck Oil Rock, just to the west of Compton Bay near Hanover Point.

The alarm was promptly given by the coastguards, and the signal rockets for the assembly of the life-boat crew were fired by Coxswain Ben Jacobs. Within a few minutes the greater part of the life-boat crew and helpers were on the spot together with the Rev. R. Leslie Morris, the secretary, Mr. Charles Seely, Mr. Hayle, coastguard officer, and other members of the life-boat committee, and a team of horses from Mr. Seely's home farm quickly followed. In a remarkably short time the boat was unhoused and being transported along the beach to the nearest point to the wreck which was free from rocks and suitable for launching, and at 6.25 a.m. – barely 25 minutes after the first signal – the boat was afloat.

A heavy sea was breaking over the unfortunate vessel, which was surrounded by surf, and on getting near enough to be heard the coxswain of the life-boat shouted his instructions to the marooned sailors, telling them to throw a rope to which the life-boat could make fast on the lee side clear of the wreckage of the bowsprit which was beating about. Two of the men, however, failing to understand, or over anxious to escape from the death which threatened them, prematurely threw themselves into the sea, and in order to rescue them several of the life-boat crew had to drop their oars, and the boat was driven away. On regaining the wreck the rescue of the remainder of the crew was quickly accomplished, and the return to shore was safely effected by 6.55 a.m.

On being brought to shore the men were regaled with hot coffee and relieved of their wet garments and supplied with blankets and with dry clothing from a small store in the boat-house. The men were afterwards entertained to breakfast at Brook House by Mr. Charles Seely, and his eldest son Mr. Charles Hilton Seely, and later in the day they were conveyed to Cowes.

18th November 1904 — Steam Dredger TRITON

Very occasionally a ship and its crew would be attended by the life-boat and leave without acknowledging the assistance given. More often, however, the heartfelt gratitude of the survivors could be embarrassing. The following letter of thanks received from the Spanish captain of the *Triton* which came ashore at Brook on 18th November 1904 exemplifies the genuine expressions of thanks that the crews received. This letter of gratitude is recorded on one of the boards in Brook Church.

> "The November 18th 1904 will be for me one of the most memorable days of my life, for on that date happened one of those terrible cases of fortune, the accident to the steam boat *Triton* on the English coast. As I manned the ship as Captain I am obliged to express to you our thanks for your splendid help in the dangers, without your help it would not have been possible to save all the crew of my ship. I am ashamed by all these amiabilities, kindness and generosities of the whole Town which prevent other accidents. Receive please by this letter the loudest expressions of our most sincere gratitude and I beg you to accept this as the smallest proof of never-ending admiration of one of the most humble Captains of the Spanish Merchant Navy".

Rafael Meanay Arias
Santona, Spain
The 18th December 1904

Launching the Susan Ashley

4th February 1916 Barque SOUVENIR

The fierce south-westerly gale which raged on the night of Thursday 3rd February 1916 resulted in the total loss of the Norwegian barque *Souvenir*. Captain Weaj, the master of the ship, and a steward named Olsen, perished. But for the magnificent services rendered by the Brook life-boat crew, in the *Susan Ashley*, the death-roll would certainly have been heavier; it was little short of miraculous that all hands did not succomb.

The *Souvenir's* crew consisted of the captain and nine others – all Norwegians. The vessel was bound to Lisbon from Tuedestrand, Norway, with a ballast of chalk. The full fury of the gale was encountered in the English Channel, and the barque was tossed about by the heavy seas which broke over her. At night the storm increased in violence, and off St. Catherine's the *Souvenir* was little better than a derelict and unmanageable. She drifted with the wind and tide, the huge waves repeatedly sweeping over her. It was a fearful night, and the unfortunate sailors were in imminent danger of drowning at any moment. They were unaware of their exact position, and to have abandoned the ship and taken to the boats would have meant certain death, for no boat could have lived in such a sea. The helpless ship drifted on in the blinding darkness of the night until a grinding noise was heard, and then it was known that she had stranded but where, those on board had no idea.

Rockets and distress signals had previously been sent up, and they were seen by watchers along the coast. The Totland Bay life-boat crew were summoned, and their steam life-boat was launched and went off in search of the ship, but did not reach her. It was about midnight when the *Souvenir* stranded, but it was not until after daybreak that the vessel was seen lying hard and fast on the dangerous Great Stag Ledge eastward of Brook Chine and about a mile from the shore. The storm had somewhat abated, but terrific seas were washing over the ship from stem to stern. Her position was hopeless, and how the crew managed to pass the long dreary hours of the night in such extremely perilous circumstances it is difficult to imagine.

On the Friday morning at about seven o'clock the Brook life-boat crew were summoned by the firing of rockets. Although several of the crew of the *Susan Ashley* – including Brigadier-General Jack Seely – were away on active war service, volunteer substitutes were found, and about an hour after the summons the life-boat was launched. However, it was found impossible to get alongside the *Souvenir,* whose decks were awash, and fierce seas were breaking right over her. The crew of the stranded barque were much relieved at the arrival on the scene of the life-boat, but their rescue was an extremely difficult matter. When it was found that the life-boat could not get alongside to bring off the crew, the Norwegians saw that their only chance of escape was to jump into the sea and trust to the life-boat to pick them up. With commendable courage, the shipwrecked crew, who had put on lifebelts, and had the further assistance of lifebuoys, jumped into the boiling sea and with the aid of lines all of them, with the exception of the captain, were rescued. Unfortunately the steward, Olsen, an elderly man some 70 years old, who was picked up in a state of collapse, succumbed to shock and exposure in the life-boat. When his body was landed every attempt to restore life by artificial respiration was made by Dr. W. J. Jolliffe, who had been summoned from Yafford, but it was unavailing. The body was placed on a stretcher and taken to the life-boat house. A silver watch, found in the deceased's pocket, had stopped at 10.15. The eight survivors, when taken out of the water, were taken to the life-boathouse and given restoratives and a meal.

The heavy buffeting by the waves caused the stranded barque speedily to break up and in a few hours all that remained of her were the fragments of her frame just projecting above the crest of the waves, and the masts, timbers, portions of cabins and other wreckage which were washed in and strewn along the shore. The ship's books were picked up, but the crew lost everything.

The crew that night consisted of Ben Jacobs, in charge, and John Cook, John Newberry, Phil Jacobs, Tom and David Hookey, James Newberry, Rowland Hayter, Walter Cotton, Harding, Holbrook, Joseph Morris and W. Hayter, the last five being volunteers. For his bravery and example of leadership that day Ben Jacobs received the R.N.L.I. Silver Medal. Shortly after this wreck Rowland Hayter succeeded Ben Jacobs as Coxswain of the Brook life-boat. Walter Cotton was later to become the first Coxswain of the Yarmouth life-boat.

By 1937 the motor life-boats stationed on the northern shores of the Island were able to deal with shipping accidents in Brighstone Bay and the Brook station was closed in March of that year. The shell of the life-boat house still remains for all to see and be reminded of the gallant work of the life-boatmen of Brook.

ATHERFIELD LIFE-BOAT STATION

RECORD OF SERVICE

Date	Casualty	Service
CATHERINE SWIFT		
June 4th 1891	S.S. *St. Margaret*	No service rendered
Jan. 31st 1892	S.S. *Eider* of Bremen	Assistance declined
February 1st 1892	S.S. *Eider* of Bremen	Loaded mail, rescued 55
February 2nd 1892	S.S. *Eider* of Bremen	Loaded mail, rescued 57
February 3rd 1892	S.S. *Eider* of Bremen	Loaded mail, silver bars
February 4th 1892	S.S. *Eider* of Bremen	Loaded mail, plate luggage
March 13th 1894	A ketch	Put boat on slipway
Feb. 19th-21st 1897	S.S. *Alcester* of Liverpool	Rescued 22
November 23rd 1898	S.S. *Huntcliff* of W. Hartlepool	Crew assembled
Feb. 15th-16th 1900	Barque *Auguste* of Brake	Rescued 18
April 15th 1901	S.S. *Albion* of Southampton	Crew assembled
January 13th 1904	S.S. *Neck* of Bremen	Crew assembled
May 28th 1904	Schooner *Dolly* of Hamburg	Rescued 5
June 3rd 1905	Barque *St. Moran* of Stavanger	Gave help
January 9th 1906	A ketch	No service rendered
October 6th 1906	Steam trawler *Dolores*	No service rendered
GEM		
March 6th 1908	A schooner	Crew assembled
Sept. 11th 1908	Yacht *Ceylon* of Colombo	No service rendered
October 24th 1908	S.S. *Cayo Solo* of London	Stood by vessel
December 27th 1909	A steamer	No service rendered
February 6th 1910	Steam trawler *Nimrod* of Ostend	No service rendered
April 19th 1910	S.S. *Diligent* of Sunderland	Stood by vessel
March 25th 1912	Unknown	Crew assembled
March 17th 1914	Unknown	Crew assembled
June 18th 1914	S.S. *Onward*	Crew assembled

STATION CLOSED NOVEMBER 1915

The first Catherine Swift *life-boat. The presence of railwaymen suggest that this photograph was taken at Portsmouth after the rail journey from London and before her crew sailed her over to the Island*

ATHERFIELD

"Atherfield, Isle of Wight – It having been considered that a life-boat placed at Atherfield, on the south-west shore of the Isle of Wight, would be the means of affording material assistance to vessels which are not unfrequently driven on to the ledges of rocks off that dangerous part of the coast, the Royal National Life-boat Institution decided last year to form a life-boat station there. Accordingly a corrugated iron boathouse has been erected on the top of the cliff – here about 76 feet high – to the face of which a series of flat skids or sleepers have been securely fastened; over these the life-boat is lowered by ropes, the incline being 1 in 3, and the distance to the beach about 240 feet. When returning to the boathouse the life-boat is hauled up by means of a powerful winch. The new boat, which was placed on its station in October last, is 31 feet long, 7½ feet wide, and rows 10 oars, double-banked. It possesses all the latest improvements, with all the other characteristics of the boats of the Institution in the way of self-righting, self-ejecting water, etc. The cost of the new boat and equipment has been defrayed from a legacy bequeathed to the Institution for the purpose by the late Mrs. Swift, of Kensington and Chale, the boat being named after her the *Catherine Swift*".

This announcement in the February 1891 edition of the R.N.L.I. journal sets the scene of the Atherfield station set up to complement the Brighstone Grange and Brook stations and so completing the original plan drawn up 12 years earlier for three stations in Brighstone Bay.

The London, Brighton and South Coast Railway Company kindly gave the life-boat free conveyance over their line from London to Portsmouth, whence it was sailed to its station on the 29th October 1890, the crew having come over especially to fetch it. When off St. Catherine's, they found a very heavy hollow sea in the race there; and the coxswain and crew afterwards reported that the boat behaved admirably and gave them every satisfaction. The boat was handled on the beach on roller chocks placed under the keel which worked well, and no difficulty was experienced in hauling the boat up the steep incline and placing it in the house. From trials it was established that a launch could be effected in about a quarter of an hour.

When the *Catherine Swift* went into service there were no less than seven Cotton brothers in the crew including William, often known as Rufus, the coxswain and David the second coxswain. These two were both experienced life-boatmen who held the Institution's silver medal for their bravery when members of the crew of the *Worcester Cadet,* on the terrible night the *Sirenia* ran ashore, as recounted on page 14.

Two years later *Catherine Swift* was to be replaced by a slightly larger boat of the same name, the cost of which was once again defrayed from the legacy of Mrs. Swift. She was to remain in service until 1906 when the new boat was to be the *Gem,* the cost of which was paid by Mrs. G. W. Moss of Liverpool. When

Gem came into service William Cotton retired as coxswain and was replaced by William White who was to remain until the station closed.

By 1914 the eroding cliffs were beginning to cause serious difficulties in launching the life-boat and the decision was taken to close the station the following year. In 1916 the life-boat house was removed and re-erected at the Institution's storeyard at Poplar and two years later was sold for £110.

During the 25 years that Atherfield operated as a life-boat station, apart from practice and demonstrations, the life-boat was launched on 39 occasions and a total of 157 lives were saved.

The following accounts of the incidents, four when lives were saved and one when great bravery was shown, exemplify the dedicated service of the men of Atherfield.

31st January 1892 S.S. EIDER

The story of the stranding of the *Eider* has been fully told on page 20 and in that story the *Catherine Swift* will be seen to have played her full part. She was the first of the three life-boats to be launched and had made the fruitless trip, when the *Eider* had first run ashore on the Sunday evening, when all she was requested to do was to bring ashore telegrams for steam tugs. She was also the last life-boat to be involved, making five trips on the Thursday after the other two had returned to their stations, ferrying ashore mail, the ships plate and passenger's luggage.

On the three intervening days the life-boatmen of Atherfield with their colleagues from Brighstone and Brook made in all 17 trips, saving 112 lives and landing much of the mail that the *Eider* had been carrying.

As with his fellow coxswains, William Cotton was to receive an inscribed gold watch from the Kaiser.

19th February 1897 S.V. ALCESTER

The full-rigged ship, *Alcester,* of Liverpool, stranded on Atherfield Ledge on the afternoon of Friday 19th February 1897, in an unusually dense fog. The vessel, of 1,597 tons, owned by Mr. R. F. Haws, of Liverpool, stated at the time to be the uncle of her sailing master, Captain Alison D. Haws, was bound from Calcutta to Hamburg with some 2,500 tons of jute. At about 4.30 in the afternoon the coastguard first observed the stranded vessel when the fog temporarily lifted, but she was quickly obscured again. It was nearly low water at the time. Captain Haws later stated that he had not sighted land after passing Start Point and had taken a deep sounding only a few minutes before the vessel struck, and that the captain had seen a steamer pass inside, between him and the Island coast, shortly before the accident.

The coxswain of the Atherfield life-boat, William Cotton, gave the signal calling his crew together, and as quickly as possible the life-boatmen put off to the hapless vessel to rescue her crew, or to render any assistance to the captain, who had previously been unsuccessful in trying to run out an anchor. The captain, expecting that his vessel would float at high water, declined assistance, intimating that he would signal for the life-boat if he needed help later. So the life-boat returned and was pulled up until Saturday.

All hopes of the vessel floating off were doomed to disappointment, as on Saturday morning she remained firmly fixed on the rocks, some 300 or 400 yards

Life-boat Gem *at Atherfield. The buoyant rubbing strake introduced into the later boats can quite clearly be seen also the wheel chocks used as no beaching carriage was provided. Also showing over the bows is the hauling out line from the cliff top*

from land, pointing straight inshore. Unsuccessful efforts were made by a tug during the early part of the day to remove the vessel, three hawsers being snapped in the attempts. It now was apparent that the ship would become a complete wreck, and with signs of an increasing ground swell and a fresh wind the position was becoming critical.

Shortly before midday distress signals were given from the *Alcester,* and the crew of the life-boat promptly responded to the second call. By about 1 o'clock they had surmounted the difficulties of launching and the life-boat was on her way. As the crew were desirous of saving their clothes and other belongings, two journeys were necessary. Twelve men were taken off on the first journey and eight on the second, together with their property. The captain and mate resolutely refused to leave the doomed ship in spite of the earnest entreaties of the coswain of the life-boat, whose fears for their safety were only too well grounded, having regard to subsequent developments.

In the evening the wind and ground-swell increased to an alarming degree, and the situation of the ill-fated ship became more and more perilous. About nine o'clock on Saturday evening the throwing-up of blue lights from the vessel indicated the captain's anxiety for the help of the life-boat. The boat's crew were again called together, but a consultation between the coxswain and the members of the local committee present resulted in a decision to postpone the launch for a time on the grounds that there was not sufficient shore room to get the boat off, having regard to the heavy ground-swell and the high water, which reached the cliff at the foot of the slipway. All were agreed that under these conditions it was impossible to get the boat off, so the crew decided to wait till early morning, the opinion of the coxswain, judging from previous experience, being that the vessel would hold together till morning.

In the meanwhile the position of Captain Haws and his mate had become alarmingly desperate, as they were driven into the rigging by the seas which were sweeping the vessel from stem to stern. In this uncomfortable quarter they had to remain the whole night. They took to the rigging about 8 p.m. on Saturday night, and only ventured down once at about 3 a.m., when the condition of things below speedily compelled them to again seek safety in the rigging until they were rescued. The captain continued at intervals during the night to throw up blue lights.

At five o'clock the life-boat gun was again fired, and the conditions now permitting a launch – the tide having turned – the life-boatmen, a number of whom had remained on the scene all night, quickly got their boat under way. It was a difficult launch, and owing to the heavy sea the life-boatmen had a tough job before them, having to pull a considerable distance out to enable them to drop down to the distressed ship. This time the captain and his mate did not wait for an entreaty to leave their ship, for the bows of the boat had hardly touched the vessel when they leapt among the life-boatmen. The boat was soon safely back to shore and thus successfully accomplished another life-saving exploit. Amongst the rescued property of the crew were two monkeys, a dog and a cat.

15th February 1900 — Barque AUGUSTE

The German three-masted iron barque *Auguste,* of Brake near Bremen, North Germany, homeward bound from Australia, was driven on to Atherfield

Ledge on the afternoon of Thursday 15th February 1900, during a severe southerly gale which was raging.

The ill-fated vessel, of 1,298 tons register, was bound from Freemantle, West Australia, to London, with a cargo of 2,000 tons of Jarrah wood. Under the command of Captain Ammerman and with a crew of 18, she stranded between 4.30 and 5 p.m. quite close to the spot where the *Eider* was wrecked.

The life-boat gun was fired promptly in response to signals of distress, and at 5.30 p.m. the Atherfield life-boat was with great difficulty launched in a heavy sea; but the terrific force of wind and tide baffled the efforts of William Cotton, by now a veteran coxswain, and his colleagues to get near the stranded vessel, and they were reluctantly compelled to temporarily abandon their rescue work.

It being thought that the Brighstone boat might stand a better chance of attempting the rescue, Mr. Charles Dabell, the secretary of the Atherfield Life-boat Committee, wired for the assistance of the Brighstone boat, which, however, it was found impossible to get out. Danger being considered to be imminent owing to the perilous position of the *Auguste,* which had swung round and was lying between 200 and 300 yards from the shore, heading out to sea. Mr. Dabell then telephoned for a tug and the Southsea life-boat.

The night wore on and the life-boatmen and the large number who had assembled on the cliff became more and more anxious for the safety of the crew, who during a portion of the night were driven into the rigging for safety, so mercilessly were the breakers dashing over the vessel. An unsuccessful attempt was made to establish communication by means of the rocket apparatus. Some time later towards midnight the wind moderated and veered from south to west, and the fury of the sea abated. The life-boatmen could be restrained no longer, and at about 2 a.m. in the morning they tried again. Still they were doomed to disappointment for a time, as the prevailing conditions rendered it impossible to successfully launch the life-boat till just before 3 a.m. Plenty of willing helpers were found among the crowd on the cliff, and the life-boatmen were cheered as their boat headed away for the *Auguste,* which was reached after a very hard pull. Captain Ammerman and his men, who had naturally suffered from exposure to the storm during the long anxious hours of the night, gratefully welcomed the assistance of the life-boat, and they were soon all safely landed without accident, amid the renewed cheers of the crowd on shore. The rescued crew, who were loud in their praise of the life-boatmen, showed their gratitude to their rescuers by giving a special and hearty round of cheers in honour of Cotton and his men. The distressed mariners were then taken to the life-boat house, where hot coffee was provided by the committee, and further comforts in the way of food and stimulants were furnished to the men.

28th May 1904 Schooner DOLLY

During dense fog in the early morning of Saturday 28th May 1904, the German-owned two-masted schooner *Dolly,* Captain Gewald, bound from Teignmouth to Bridgeness, Scotland, with a cargo of china clay, stranded on Atherfield Ledge. The vessel was so damaged that it had to be abandoned by the captain and crew of five some hours later, the crew being taken off in the Atherfield life-boat.

At the same time a large steamer, thought to be a liner, but the name of which was not ascertained owing to the darkness and fog, passed inside the *Dolly* and also stranded, but apparently missed the rocks and thus had an almost miraculous escape from wreck as she refloated and was able to proceed on her journey after the rising of the tide.

The *Dolly* had left Teignmouth in dirty weather on the previous Friday morning and Captain Gewald ran into the first dense fog he had experienced off the Isle of Wight. He later said that the passing of the large steamer between him and the Island coast gave him confidence that he was keeping a correct course, but soon after the *Dolly* ran on to the ledge of rocks off Atherfield. It was evident that the *Dolly* was badly damaged, as water began to rise in the hold, and in spite of all that could be done the vessel remained fast. The steamer which had stranded some distance away refloated in about three hours.

Chief Officer Hirons, of the coastguards, first observed from shore the perilous position of the vessels, and about 1.15 a.m. he fired the rocket and the life-boat crew were immediately summoned. The life-boatmen promptly responded and in an hour and a half the *Catherine Swift* was launched. The steamer had by this time, fortunately, got away. The life-boat stood by the *Dolly* during the morning until the captain decided to leave the ship, and the captain and crew were then taken off by the life-boat and safely landed about midday.

During the morning the tug *Irishman,* from Cowes, arrived and attempted to tow the vessel off, but the two-rope broke twice and the effort had to be abandoned.

Captain Gewald and his crew were shown every kindness on shore by the coastguard and others, and they were very grateful for the arrangements made for their comfort.

3rd June 1905 — Barque ST. MORAN

About three o'clock on the morning of Saturday 3rd June 1905, in dense fog, the Norwegian Barque *St. Moran,* 1,500 tons register, bound from Buenos Aires to London with a cargo of wheat and oats, stranded on Side Pit Ledge at Atherfield.

At 3.15 a.m. the life-boat gun was fired by the coastguard, and the life-boatmen under Coxswain Walter White promptly responded. The life-boat put off, as it was feared that with a heavy ground sea running the *St. Moran* might be driven on to the rocks. The stranded barque, with a crew of 17, was fortunately westward of the highest ledges where enormous breakers were rolling in. The life-boatmen succeeded in working the vessel off some distance by means of a kedge and almost floated her, but as there was no other kedge available they had to wait the arrival of a Southampton tug which safely towed the vessel off at high water about 10.00 a.m.

The *St. Moran* then proceeded westward, reportedly totally ignoring the valued assistance given to the ship by the life-boatmen, but for whose efforts she would probably have been wrecked.

BRIGHSTONE BAY 1985

The change from sail to steam that took place at the end of the 19th and the beginning of the 20th centuries greatly reduced the risk of disasters on this part of the coast. Improved navigational aids and the fact that for a steamship a lee shore, so often the cause of sailing ships disasters, no longer held the same fear. Added to this the difficulties in launching a heavy beach boat despite the brave and willing efforts of the crew and beach team and the continuing expense of maintaining the access were all factors that led to the R.N.L.I. deciding that the Brighstone and Atherfield stations should be closed. The final meeting of the local committee took place on 12th July 1915 when the decision of the R.N.L.I. was announced. At that meeting it was resolved that the many thanks of the Committee should be accorded to the crew and helpers for the efficient way in which they had carried out their services in the course of life-saving.

Brook was to remain as a life-boat station until 1937 when the advent of motor lifeboats at Totland and Yarmouth caused this last station in Brighstone Bay to be closed.

Writing in the Parish Magazine at the time the Brighstone station closed, the Rector referred to the closing of a service which had played a real and honourable part in the life of the village since first a life-boat was placed at Brighstone in 1860. He also referred to the excellent work done by the crews under the leadership of three first-rate coxswains, Mr. James Buckett, Mr. Moses Munt and Mr. James Cotton, the last still at his post to the end.

Of the life-boatmen themselves. James Buckett, born in Trafalgar year, the son of an Isle of Wight shepherd, whose career had started as a fisherman, had included a spell of five years in the Navy and 18 years as coxswain of the Brighstone boat, lived to a great age. It is recorded that on Easter Day 1898 he and Daniel Creeth the sexton both in their 93rd year knelt at the altar of St. Mary's Church for holy communion. He was to die two years later six months short of his 95th birthday. James Cotton, the last coxswain at Brighstone, also lived to a great age, dying on 9th June 1928.

Although the village of Brook remains mainly unchanged much of Brighstone today would be unrecognisable to the life-boatmen of 100 years ago. With a population approaching 1,500 the village has over the past 20 years become one of the fastest growing residential areas in the Island. There are, however, still living in the village the descendents of those crewmen, Fisher's Cottage in the appropriately named Munt's Lane is still in the possession of the family of the famous coxswain and the *Sirenia* is recalled in the name of one of the roads in a new housing estate.

In St. Mary's Churchyard the graves of the three victims of the *Sirenia* disaster are carefully tended, the headstones are now lichen covered and the lettering is becoming difficult to decipher but the inscription can still be read: "Their deeds will be told while the ocean roars o'er the Ledges of Atherfield".

Sirenia Close, Brighstone 1983

Grange Chine 1984

They will stand forever as a constant reminder of that terrible night. Most years a well attended life-boat service is held in the Parish Church when flowers are placed on the graves in the churchyard, the life-boatmen of old are remembered and the collection goes to the R.N.L.I. for the life-boatmen of today.

At the St. Catherine's Quay Exhibition at Blackgang can be seen the model of the *Worcester Cadet* presented to Captain Henderson-Smith by the R.N.L.I. and a model of the second *Catherine Swift*. Around them are pictures and mementos of the work of the life-boatmen of Brighstone Bay. Of particular interest is the leather belt and brass token mentioned on page 4 as authority for a life-boat to be launched.

At Atherfield Point nothing remains to mark where the life-boat shed stood but at Brook the shell of the stone building still stands at the head of the chine.

At Grange Chine little remains of the life-boat house. Following the closure of the station the building became surplus to requirements and between the two World Wars it saw service as a beach cafe. Early in the Second World War it was demolished by a high explosive bomb and all that remained were the foundations. As the photograph on page 2 shows little now remains of these foundations and before very much longer the falling cliffs, so prevalent in this part of the Island, will finally remove all trace of the building.

Standing on the cliff at Atherfield or at the ends of the chines at Brook and Brighstone Grange and looking out over the beaches and ledges on a fine summer's day it is difficult to imagine the scenes of excitement which took place here 100 years ago. But with a gale blowing and a high sea running it is quite possible to envisage the hustle and bustle of the helpers, to hear the cries of the crews of the wrecked ships, and to visualise the *Eider* making her stately pirouette before finally settling on the rocks.

If there were to be such a shipwreck today the Yarmouth life-boat could be round in 30 minutes and the Bembridge boat would not be far behind her, but the work of the old life-boatmen is not forgotten – the Munts, the Cottons, the Bucketts, the Jacobs, the Downers, the Salters and all the brave life-boatmen of Brighstone Bay – we remember you.

"O hear us when we cry to thee
for those in peril on the sea".

BIBLIOGRAPHY

Back of the Wight	Fred Mew	Isle of Wight County Press
Isle of Wight Life-boats	G. E. Cotton	R.N.L.I. 1981
Adventure	Major General J. E. B. Seely	Hodder & Stoughton
Fear, and be slain	Major-General J. E. B. Seely	Hodder & Stoughton 1931
For Ever England	Major-General J. E. B. Seely	Hodder & Stoughton 1932
Launch	Major-General J. E. B. Seely	Hodder & Stoughton 1938
Let Not the Deep	A. D. Farr	Impulse Books 1973
Life-boat in Danger's Hour	Patrick Haworth	Hamlyn 1981
The Life-boat in Verse	Sir John Cunningham and Charles Vince	Hodder & Stoughton 1938
The Life-boat Service	Oliver Warner	Cassell & Co. 1974
	R.N.L.I. Archives	
	Brighstone Parish Records	
	Illustrated London News	
	Isle of Wight County Press	
	Isle of Wight Observer	
	Punch	
	Thames Nautical Training College H.M.S. Worcester Annual Reports	
	The Marine Engineer	

APPENDIX

SPECIFICATIONS AND DETAILS OF LIFE-BOATS

To give an indication of the specification of the boats and their launching carriages and complements the following are the descriptions of those of *Worcester Cadet*.

Boat:— Diagonal Honduras mahogany planking gunwhale to gunwhale. Self-righting compartments bow and stern, self-righting tanks fitted 1888.

Complement:— Bowman, ten oarsmen and two coxswain.

Carriage:— Forward bogie two small wheels. Connecting shaft for pair of horses. Two six-foot diameter main centre wheels. Wheels with wide iron rims. Tipping plates to main wheels for soft sand conditions. Drag chains and securing ropes.

Life-boat	*Official Number*	*Builders*	*Cost* £	*s.*	*d.*	*Length*	*Breadth*	*In Service*	*Withdrawn*
BRIGHSTONE GRANGE									
Rescue	—	Forrestt Peake, Limehouse	143	0	0	30ft. 0in.	7ft. 2in.	13.8.1860	25.7.1866
Rescue	—	Forrestt & Son, Limehouse				30ft. 0in.	7ft. 2in.	25.7.1866	11.8.1880
Worcester Cadet	226	Woolfe & Sons, Shadwell, London	363	0	0	34ft. 2in.	8ft. 1½in.	11.8.1880	17.10.1892
Joe Jarman	328	W. J. Ellis	379	3	6	34ft. 1in.	8ft. 3⅜in.	17.10.1892	30.5.1907
The Brothers Brickwood	574	Thames Iron Works	871	4	4	35ft. 0in.	8ft. 6in.	30.5.1907	7.10.1915
BROOK									
Dauntless	—	Forrestt Peake, Limehouse	148	9	6	30ft. 0in.	7ft. 0in.	13.8.1860	5.1867
George and Anne	—	Forrestt & Son, Limehouse	272	11	6	32ft. 0in.	7ft. 6in.	5.1867	30.1.1888
William Slaney Lewis	144	Hansen & Sons	347	9	10	34ft. 1¼in.	7ft. 6in.	30.1.1888	30.5.1907
Susan Ashley	579	Thames Iron Works	843	10	1	35ft. 0in.	8ft. 6in.	5.9.1907	31.3.1937
ATHERFIELD									
Catherine Swift	278	Waterman Bros.	252	0	0	31ft. 0in.	7ft. 3in.	27.10.1890	29.12.1892
Catherine Swift	354	Woolfe & Sons, Shadwell, London	458	0	0	34ft. 2in.	7ft. 6½in.	29.12.1892	3.12.1906
Gem	568	Thames Iron Works	830	10	10	35ft. 0in.	8ft. 6in.	3.12.1906	2.12.1915

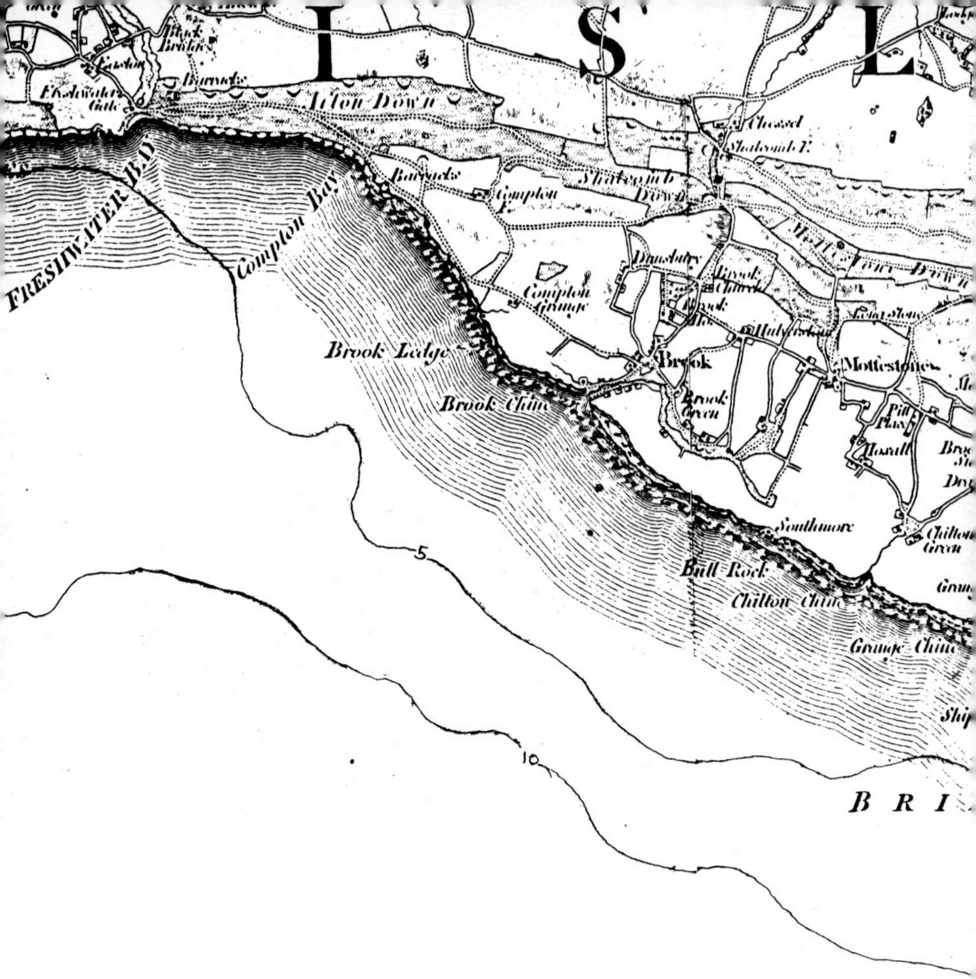

I
S
L
Afton Down
FRESHWATER BAY
Compton Bay
Barracks
Compton F.
Compton Grange
Shalcomb Down
Dunsbury
Brook
Brook Ledge
Brook Chine
Brook Green
Mottestone
Southmore
Bull Rock
Chilton Chine
Grange Chine
5
10
B R I